Implementing Software Defined Radio

Eugene Grayver

Implementing Software Defined Radio

 Springer

Eugene Grayver
Walnut Ave 2317
Manhattan Beach
CA 90266
USA

ISBN 978-1-4419-9331-1   ISBN 978-1-4419-9332-8   (eBook)
DOI 10.1007/978-1-4419-9332-8
Springer New York Heidelberg Dordrecht London

Library of Congress Control Number: 2012939042

Printed on acid-free paper

Springer is part of Springer Science+Business Media (www.springer.com)

To my parents, who have taught me how to be happy, and that there's more to life than science. And to my grandparents, who value science above all and have inspired and guided me in my career

Preface

A search for 'Software Defined Radio' on Amazon.com at the end of 2010 shows that almost 50 books have been written on the subject. The earliest book was published in 2000 and a steady stream of new titles has been coming out since. So why do I think that yet another book is warranted?

SDR is now a mature field, but most books on the subject treat it as a new technology and approach SDR from a theoretical perspective. This book brings SDR down to earth by taking a very practical approach. The target audience is practicing engineers and graduate students using SDR as a tool rather than an end unto itself, as well as technical managers overseeing development of SDR. In general, SDR is a very practical field—there just isn't very much theory that is unique to flexible radios versus single function radios.[1] However, the devil is in the details… a designer of an SDR is faced with a myriad of choices and tradeoffs and may not be aware of many of them. In this book I cover, at least superficially, most of these choices. Entire books can be devoted to subjects treated in a few paragraphs[2] below (e.g. wideband antennas). This book is written to be consulted at the start of an SDR development project to help the designers pin down the hardware architecture. Most of the architectures described below are based on actual radios developed by the author and his colleagues. Having built, debugged, and tested the different radios; I will highlight some of the non-obvious pitfalls and hopefully save the reader countless hours. One of my primary job responsibilities is oversight of SDR development by many government contractors. The lessons learned from dozens of successful and less than successful projects are sprinkled throughout this book, mostly in the footnotes.

Not every section of this book addresses SDR specifically. The sections on design flow and hardware architectures are equally applicable to many other digital designs. This book is meant to be at least somewhat standalone since a

---

[1] Cognitive radio, which is based on flexible radio technology, does have a significant theoretical foundation.

[2] The reader is encouraged to consult fundamental texts referenced throughout.

practicing engineer may not have access to, or the time to read, a shelf full of communications theory books. I will therefore guide the reader through a whirl-wind tour of wireless communications in Appendix A.[3] The necessarily superficial overview is not meant to replace a good book on communications [1,2] and the reader is assumed to be familiar with the subject.

The author does not endorse any products mentioned in the book.

---

[3] The reader is encouraged to at least skim through it to become familiar with terminology and nomenclature used in this book.

Acknowledgments

Most of the ideas in this book come from the author's experiences at two companies—a small startup and a large government lab. I am fortunate to be working with a truly *nulli secundus* team of engineers. Many of the tradeoffs described in this text have been argued for hours during impromptu hallway meetings. The nature of our work at a government lab requires every engineer to see the big picture and develop expertise in a wide range of fields. Everyone acknowledged below can move effortlessly between algorithm, software, and hardware development and therefore appreciate the coupling between the disciplines. This book would not be possible without the core SDR team: David Kun, Eric McDonald, Ryan Speelman, Eudean Sun, and Alexander Utter. I greatly appreciate the invaluable advice and heated discussions with Konstantin Tarasov, Esteban Valles, Raghavendra Prabhu, and Philip Dafesh. The seeds of this book were planted years ago in discussions with my fellow graduate students and later colleagues, Ahmed ElTawil and Jean Francois Frigon.

I am grateful to my twin brother for distracting me and keeping me sane. Thanks are also due to my lovely and talented wife for editing this text and putting up with all the lost vacation days.

Contents

1   What is a Radio? 1

2   What Is a Software-Defined Radio? 5

3   Why SDR? 9
   3.1   Adaptive Coding and Modulation 10
       3.1.1   ACM Implementation Considerations 16
   3.2   Dynamic Bandwidth and Resource Allocation 17
   3.3   Hierarchical Cellular Network 19
   3.4   Cognitive Radio 20
   3.5   Green Radio 25
   3.6   When Things go Really Wrong 26
       3.6.1   Unexpected Channel Conditions 27
       3.6.2   Hardware Failure 27
       3.6.3   Unexpected Interference 28
   3.7   ACM Case Study 29
       3.7.1   Radio and Link Emulation 30
       3.7.2   Cross-Layer Error Mitigation 32

4   Disadvantages of SDR. 37
   4.1   Cost and Power 37
   4.2   Complexity 38
   4.3   Limited Scope 40

5   Signal Processing Devices 43
   5.1   General Purpose Processors 43
   5.2   Digital Signal Processors 44
   5.3   Field Programmable Gate Arrays 44

5.4   Specialized Processing Units   47
        5.4.1   Tilera Tile Processor.   49
5.5   Application-Specific Integrated Circuits   51
5.6   Hybrid Solutions   51
5.7   Choosing a DSP Solution   52

6   **Signal Processing Architectures.**   55
    6.1   GPP-Based SDR.   55
            6.1.1   Nonrealtime Radios   58
            6.1.2   High-Throughput GPP-Based SDR   60
    6.2   FPGA-Based SDR   60
            6.2.1   Separate Configurations.   61
            6.2.2   Multi-Waveform Configuration   61
            6.2.3   Partial Reconfiguration   62
    6.3   Host Interface   68
            6.3.1   Memory-Mapped Interface to Hardware   69
            6.3.2   Packet Interface   73
    6.4   Architecture for FPGA-Based SDR.   73
            6.4.1   Configuration.   73
            6.4.2   Data Flow   75
            6.4.3   Advanced Bus Architectures   78
            6.4.4   Parallelizing for Higher Throughput   80
    6.5   Hybrid and Multi-FPGA Architectures   81
    6.6   Hardware Acceleration   83
            6.6.1   Software Considerations   84
            6.6.2   Multiple HA and Resource Sharing   89
    6.7   Multi-Channel SDR   92

7   **SDR Standardization**   97
    7.1   Software Communications Architecture and JTRS   97
            7.1.1   SCA Background   98
            7.1.2   Controlling the Waveform in SCA   103
            7.1.3   SCA APIs   104
    7.2   STRS   107
    7.3   Physical Layer Description   109
            7.3.1   Use Cases   111
            7.3.2   Development Approach.   111
            7.3.3   A Configuration Fragment.   113
            7.3.4   Configuration and Reporting XML.   115
            7.3.5   Interpreters for Hardware-Centric Radios.   116
            7.3.6   Interpreters for Software-Centric Radios   116
            7.3.7   Example   118

7.4   Data Formats   118
    7.4.1   VITA Radio Transport (VITA 49, VRT)   118
    7.4.2   Digital RF (digRF)   125
    7.4.3   SDDS   125
    7.4.4   Open Base Station Architecture Initiative   127
    7.4.5   Common Public Radio Interface   128

8   Software-Centric SDR Platforms   131
8.1   GNURadio   131
    8.1.1   Signal Processing Blocks   132
    8.1.2   Scheduler   135
    8.1.3   Basic GR Development Flow   136
    8.1.4   Case Study: Low Cost Receiver
           for Weather Satellites   137
8.2   Open-Source SCA Implementation: Embedded   140
8.3   Other All-Software Radio Frameworks   143
    8.3.1   Microsoft Research Software Radio (Sora)   143
8.4   Front End for Software Radio   144
    8.4.1   Sound-Card Front Ends   145
    8.4.2   Universal Software Radio Peripheral   145
    8.4.3   SDR Front Ends for Navigation Applications   149
    8.4.4   Network-Based Front Ends   149

9   Radio Frequency Front End Architectures   151
9.1   Transmitter RF Architectures   151
    9.1.1   Direct RF Synthesis   152
    9.1.2   Zero-IF Upconversion   154
    9.1.3   Direct-IF Upconversion   155
    9.1.4   Super Heterodyne Upconversion   157
9.2   Receiver RF Front End Architectures   157
    9.2.1   Six-Port Microwave Networks   158

10   State-of-the-Art SDR Components   159
10.1   SDR Using Test Equipment   159
    10.1.1   Transmitter   160
    10.1.2   Receiver   161
    10.1.3   Practical Considerations   163
10.2   SDR Using COTS Components   165
    10.2.1   Highly Integrated Solutions   165
    10.2.2   Non-Integrated Solutions   166
    10.2.3   Analog-to-Digital Converters (ADCs)   167
    10.2.4   Digital to Analog Converters (DACs)   171

10.3   Exotic SDR Components............................... 171
10.4   Tunable Filters.................................... 173
10.5   Flexible Antennas.................................. 178

11 Development Tools and Flows........................... 183
11.1   Requirements Capture............................. 183
11.2   System Simulation 186
11.3   Firmware Development............................. 188
       11.3.1   Electronic System Level Design................ 188
       11.3.2   Block-Based System Design 190
       11.3.3   Final Implementation 192
11.4   Software Development 193
       11.4.1   Real-Time Versus Non-Real-Time Software 193
       11.4.2   Optimization 195
       11.4.3   Automatic Code Generation.................. 196

12   Conclusion.. 199

Appendix A: An Introduction to Communications Theory 201

Appendix B: Recommended Test Equipment 243

Appendix C: Sample XML Files for an SCA Radio 245

Bibliography.. 253

Index 265

Abbreviations

ACM	Adaptive coding and modulation
ADC	Analog to digital converter
AMC	See ACM
ASIC	Application specific integrated circuit. Usually refers to a single-function microchip, as opposed to a GPP or an FPGA
AWGN	Additive white Gaussian noise
CMI	Coding and modulation information. A set of variables describing the mode chosen by an ACM algorithm
CORBA	Common object request broker architecture. A standard that enables software components written in multiple computer languages and running on multiple computers to work together. Used in SCA/JTRS compliant radios
COTS	Commercial off-the-shelf. Refers to items that do not require in-house development
CR	Cognitive radio. A radio that automatically adapts communications parameters based on observed environment
DAC	Digital to analog converter
dB	Decibels. Usually defined as $10 \log_{10}(A/B)$ if the units of A and B are power, and $20 \log_{10}(A/B)$ if the units of A and B are voltage
DCM	See ACM
DDFS	Direct digital frequency synthesizer. A digital circuit that generates a sinusoid at a programmable frequency
DDS	See DDFS
DSP	Digital signal processing
DSSS	Direct sequence spread spectrum. A modulation that uses a high-rate pseudorandom sequence to increase the bandwidth of the signal
E_b/N_0	Ratio of the signal energy used to transmit one bit to the noise level. This metric is closely related to SNR
ENOB	Effective number of bits. Used to characterize performance of an ADC/DAC. Always less than the nominal number of bits
FEC	Forward error correction

FIFO	First-in-first-out buffer. A fixed-length queue
FIR	Finite impulse response. Usually refers to a purely feed-forward filter with a finite number of coefficients
FLOP	Floating point operations per second. A metric used to computer signal processing capabilities of different devices. Usually prefixed with a scale multiplier (e.g. GFLOPS for 10^9 FLOPS)
GbE	Gigabit Ethernet. 10 GbE is 10 gigabit Ethernet
GPGPU	General purpose GPU. A GPU with additional features to allow its use for non-graphics applications such as scientific computing
GPP	General purpose processor. Major vendors include Intel, ARM, AMD
GPU	Graphics processing unit. A chip (usually in a PC) dedicated to generating graphical output. A GPU is a key part of a video card
GUI	Graphical user interface
HPC	High performance computing. Also known as scientific computing. Many of the tools and techniques used in HPC are applicable to SDR development
Hz	Unit of frequency, corresponding to one period per second (e.g. a 10 Hz sine wave has 10 periods in 1 s)
I	The in-phase (real) component of a complex value
IDL	Inteface description language. Provides a way to describe a software component's interface in a language-neutral way. Used extensively in SCA-compliant radios
IF	Intermediate frequency. Refers to either an RF carrier in a super-heterodyne front end, or to digitized samples at the ADC/DAC
IIR	Infinite impulse response. Usually refers to a filter with feedback
IQ	Inphase/quadrature (complex valued). Often refers to baseband signals (vs. IF signals)
LDPC	Low density parity check code. A modern forward error correction block code. Provides coding gain similar to that of a Turbo code
LSB	Least significant bit
MAC	1. Generic multiply and accumulate operation. See FLOP 2. Medium access control (network layer)
MIMO	Multiple input multiple output. A communications system employing multiple antennas at the transmitter and/or receiver
NVRAM	Non-volatile memory. Memory that retains its data without power (e.g. FLASH, EEPROM)
OCP	Open Core Protocol defines an interface for on-chip subsystem communications
OFDM	Orthogonal frequency division multiplexing. A modulation scheme that uses many tightly packed narrowband carriers
OFDMA	Orthogonal frequency division multiple access. Multi-user version of the OFDM modulation. Multiple access is achieved in OFDMA by assigning subsets of subcarriers to individual users

OMG	Object management group. A consortium focused on providing standards for modeling software systems. Manages IDL and UML standards
OP	Operation per second. See FLOPS
POSIX	Portable Operating System Interface. A family of standards specified by the IEEE for maintaining compatibility between operating systems
PR	Partial reconfiguration. A technique to change the functionality of part of the FPGA without disturbing operation of other parts of the same FPGA
Q	The quadrature (imaginary) component of a complex value
RF	Radio frequency
RFFE	RF front end – components of the radio from the antenna to the ADC/ DAC
RX, Rx	Receive or receiver
SCA	Software communications architecture. Used in SCA/JTRS compliant radios
SDR	Software defined radio
SDRAM	Synchronous dynamic memory. Fast, inexpensive memory volatile memory
SF	Spreading factor. Number of chips per symbol in a DSSS system
SIMD	Single instruction multiple data. A programming paradigm and corresponding hardware units that allow the same operation to be performed on many values. SIMD instructions can be executed faster because there is no overhead associated with getting a new instruction for every data value
SISO	Single input single output. A conventional communications system employing one antenna at the transmitter and one at the receiver
SNDR	Signal to noise and distortion ratio. Similar to SNR, but takes non-noise (e.g. spurs) distortions into account
SNMP	Simple Network Management Protocol defines a standard method for managing devices on IP networks. Commands can be sent to devices to configure and query settings
SNR	Signal to noise ratio. Usually expressed in dB
sps	Samples per second. Frequently used with a modifier (e.g. Gsps means 10^9 samples per second)
SRAM	Static random access memory. Memory content is lost when power is removed. Data is accessed asynchronous. SRAM-based FPGAs use this memory to store configuration
TX, Tx	Transmit or transmitter
UML	Unified modeling language. A graphical language provides templates to create visual models of object-oriented software systems. Used for both requirements capture and automatic code generation
VLIW	Very long instruction word. A microprocessor architecture that executes operations in parallel based on a fixed schedule determined when programs are compiled

VQI   Video quality indicator. A quantitative metric related to the perceived
          quality of a video stream. Range of 0 to 100
VRT   VITA radio transport, VITA-49. A standard for exchanging data in a
          distributed radio system
VSWR   Voltage standing wave ratio. Used as measure of efficiency for
          antennas. An ideal antenna has VSWR = 1:1, and larger numbers
          indicate imperfect signal transmission
w.r.t.   With respect to

Chapter 1
What is a Radio?

Before discussing software-defined radio, we need to define: what is a *radio*. For the purposes of this book, a radio is any device used to exchange digital[1] information between point A and point B. This definition is somewhat broader than the standard concept of a radio in which it includes both wired and wireless communications. In fact, most of the concepts that will be covered here are equally applicable to both types of communications links. In most cases no distinction is made, since it is obvious that, for example, a discussion of antennas is only applicable to wireless links. A top-level diagram of a generic radio[2] is shown in Fig. 1.1.

In the case of a receiver, the signal flow is from left to right.

- *Antenna*. Electromagnetic waves impinge on the antenna and are converted into an electrical signal. The antenna frequently determines the overall performance of the radio and is one of the most difficult components to make both efficient and adaptable. The antenna can vary in complexity from a single piece of metal (e.g., a dipole) to a sophisticated array of multiple elements. In the past, an antenna was a passive component, and any adaptation was performed after the wave had been converted into an electrical signal. Some of the latest research has enabled the mechanical structure of the antenna itself to change in response to channel conditions. Active and adaptive antennas will be discussed in Sect. 10.5.
- *Radio frequency (RF) front end*. The electrical signal from the antenna is typically conditioned by a RF front end (RFFE). The electrical signal is typically extremely weak[3] and can be corrupted by even low levels of noise. The ambient noise from the antenna must be filtered out and the signal amplified before it can

---

[1] Analog radios are being rapidly phased out by their digital counterparts and will not be covered in this book. Even the two major holdouts, AM and FM radios, are slowly being converted to digital.

[2] An entirely different kind of radio is described in [279]. All of the functionality is (incredibly) implemented in a single nanotube.

[3] A received signal power of -100 dBm (~ 2 μV) is expected by small wireless devices, while -160 dBm (~ 2 nV) received power is common for space communications.

E. Grayver, *Implementing Software Defined Radio*,
DOI: 10.1007/978-1-4419-9332-8_1, © Springer Science+Business Media New York 2013

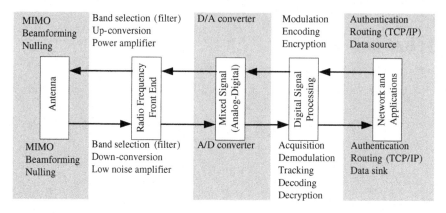

Fig. 1.1 Conceptual block diagram of a generic radio (transmit path on the *top*, receive at the *bottom*)

be processed further. The RF front end determines the signal-to-noise ratio (SNR) with which the rest of the radio has to work. A classical RF front end consists of a filter, low noise amplifier, and a mixer to convert the signal from radio frequency to a lower frequency. It is very challenging to design an RF front end that is both efficient and flexible. These two elements are currently the limiting factors in the development of truly universal radios. Classical and alternative RF front ends will be discussed in Chap. 9.

- *Mixed signal converters.* The amplified electrical signal at the output of the RF front end may be digitized for further processing. Exponential improvement in the speed and capacity of digital signal processing (DSP) has made DSP the obvious choice[4] for the development of flexible radios. A mixed signal circuit (analog to digital converter) creates a digital representation of the received signal. The digital representation necessarily loses some information due to finite precision and sample rate. Allocation of resources (power, size, cost, etc.) between the RF front end and the mixed signal converters is a key tradeoff in SDR. The mixed signal components will be discussed in Sect. 10.2.

- *Digital signal processing.* DSP is applied to extract the information contained in the digitized electrical signal into user data. The DSP section is (somewhat arbitrarily) defined as receiving the digitized samples from the mixed signal subsection and outputting decoded data bits. The decoded data bits are typically not the final output required by the user and must still be translated into data packets, voice, video, etc. A multitude of options are available to the radio designer for implementing the signal processing. A significant portion of this

---

[4] Papers presenting very creative and elegant approaches to analog implementation of traditional digital blocks are always being published (e.g., [274]), but are more exercises in design rather than practical solutions. Exceptions include niche areas such as ultra-low power and ultra-high rate radios.

book is devoted to describing various DSP options (Chap. 5) and architectures
(Chap. 6). The exact functions implemented in the DSP section are determined
by the details of the waveform,[5] but in general include:

- *Signal acquisition*. The transmitter, channel, and the receiver each introduce
  offsets between the expected signal parameters and what is actually received.[6]
  The receiver must first *acquire* the transmitted signal, i.e., determine those
  offsets (see Sect. A.9.8).
- *Demodulation*. The signal must be demodulated to map the received signal
  levels to the transmitted symbols.[7] Since the offsets acquired previously may
  change with time, the demodulator may need to *track* the signal to continuously update the offsets.
- *Decoding*. Forward error correction algorithms used in most modern radios
  add overhead bits (parity) to the user data. Decoding uses these parity bits to
  correct bits that were corrupted by noise and interference.
- *Decryption*. Many civilian and most military radios rely on encryption to
  ensure that only the intended recipient of the data can access it. Decryption is
  the final step before usable data bits are available.

- *Network and Applications*. With the exception of a few very simple, point-to-
  point radios, most modern radios interface to a network or an application. In the
  past, the network and applications were designed completely separately from the
  radio. SDR often requires tight coupling between the radio and the higher layers
  (see Table 2.1). This interaction will be discussed in Sect. 3.7 and mentioned
  throughout the book.

---

[5] Only receiver functionality is described here for brevity. Most functions have a corresponding
equivalent in the transmitter.

[6] Typical offsets include differences in RF carrier frequency, transmission time, and signal
amplitude.

[7] A simple demodulator maps a received value of +1 V to a data bit of '1' and a received −1 V to
a data bit of '0'. Modern demodulators are significantly more complex.

Chapter 2
What Is a Software-Defined Radio?

Historically, radios have been designed to process a specific waveform.[1] Single-function, application-specific radios that operate in a known, fixed environment are easy to optimize for performance, size, and power consumption. At first glance most radios appear to be single function—a first-generation cellular phone sends your voice, while a WiFi base station connects you to the Internet. Upon closer inspection, both of these devices are actually quite flexible and support different waveforms. Looking at all the radio devices in my house, only the garage door opener and the car key fob seem to be truly fixed. With this introduction, clearly a software-defined radio's main characteristic is its ability to support different waveforms.

The definition from wireless innovation forum (formerly SDR forum) states [3]:

> A software-defined radio is a radio in which some or all of the physical layer functions are software defined.

Let us examine each term individually:

- The term physical layer requires a bit of background. Seven different *layers* are defined by the Open Systems Interconnection (OSI) model [4], shown in Table 2.1.

  This model is a way of subdividing a communications system into smaller parts called layers. A layer is a collection of conceptually similar functions that provide services to the layer above it and receives services from the layer below it. The layer consisting of the first four blocks in Fig. 1.1 is known as the *physical* layer.

- The broad implication of the term *software defined* is that different waveforms can be supported by modifying the software or firmware but not changing the hardware.

---

[1] The term *waveform* refers to a signal with specific values for all the parameters (e.g., carrier frequency, data rate, modulation, coding, etc).

E. Grayver, *Implementing Software Defined Radio*,
DOI: 10.1007/978-1-4419-9332-8_2, © Springer Science+Business Media New York 2013

Table 2.1 OSI seven-layer model

	Data unit	#	Name	Function
Host layers	Data	7	Application	Network process to application
		6	Presentation	Data representation and encryption
		5	Session	Interhost communication
	Segment	4	Transport	End-to-end connections and reliability
Media layers	Packet	3	Network	Path determination, logical addressing
	Frame	2	Data Link	Physical addressing
	Bit	1	Physical	Media, signal, and binary transmission

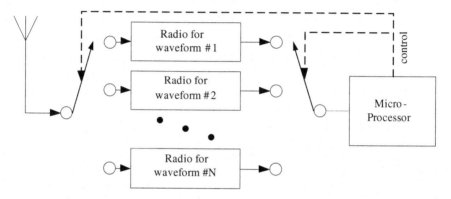

Fig. 2.1 Basic software-controlled radio

According to the strictest interpretation of the definition, most radios are not software *defined* but rather software *controlled*. For example, a modern cellular phone may support both GSM (2G) and WCDMA (3G) standards. Since the user is not required to flip a switch or plug in a separate module to access each network, the standard selection is controlled by software running on the phone. This defines the phone as a software-controlled radio. A conceptual block diagram of such a radio is shown in Fig. 2.1. Software running on a microcontroller selects one of the single-function radios available to it.

A simple thought experiment shows that the definition of a true SDR is not quite as black and white as it appears. What if instead of selecting from a set of the complete radios, the software could select one of the building blocks shown in Fig. 1.1? For example, the software would connect a particular demodulation block to a decoder block. The next logical step calls for the software to configure details of the demodulator. For example, it could choose to demodulate QPSK or 8PSK symbols. Taking this progression to an extreme, the software could define interconnect between building blocks as simple as registers, logic gates, and multipliers, thereby realizing any signal processing algorithm. Somewhere in this evolution, the software-controlled radio became a software-defined radio.

(a) **(b)**

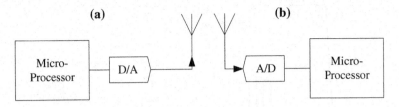

Fig. 2.2 Ideal software-defined radio: (**a**) transmitter, (**b**) receiver

The key, albeit subtle, difference is that a software-*controlled* radio is limited to functionality explicitly included by the designers, whereas a software-*defined* radio may be reprogrammed for functionality that was never anticipated.

The ideal software-defined radio is shown in Fig. 2.2. The user data is mapped to the desired waveform in the microprocessor. The digital samples are then converted directly into an RF signal and sent to the antenna. The transmitted signal enters the receiver at the antenna, is sampled and digitized, and finally processed in real time by a general purpose processor. Note that the ideal SDR in contrast with Fig. 1.1, does not have an RFFE and a microprocessor has replaced the generic DSP block. The ideal SDR hardware should support any waveform at any carrier frequency and any bandwidth.

So, what challenges must be overcome to achieve this elegant radio architecture?

- Most antennas are mechanical structures and are difficult to tune dynamically. An ideal SDR should not limit the carrier frequency or bandwidth of the waveform. The antenna should be able to capture electromagnetic waves from very low frequencies (e.g., <1 MHz) to very high frequencies (e.g. >60 GHz[2]). Section 10.5 detail the challenge of designing such an antenna. Such a wideband antenna, if available, places high demands on the RF front end (RFFE) and the digitizer.
- Selection of the desired signal and rejection of interferers (channel selection) is a key feature of the RFFE. However, the antenna and filter(s) required to implement the channel selection are usually electromechanical structures and are difficult to tune dynamically (see Sect. 10.4).
- Without an RF front end to select the band of interest, the entire band must be digitized. Following Nyquist's criterion, the signal must be sampled twice at the maximum frequency (e.g., 2 × 60 GHz). Capabilities of currently available A/D converters are discussed in Sect. 10.2.3, and are nowhere close to 120 GHz.
- The captured spectrum contains the signal of interest and a multitude of other signals, as shown in Fig. 2.3. Interfering signals can be much stronger than the signal of interest.[3] A power difference of 120 dB is not unreasonable. The

---

[2] 60 GHz is the highest frequency used for terrestrial commercial communications in 2011.

[3] Consider a practical example of a cell phone transmitting at +30 dBm while receiving a −60 dBm signal.

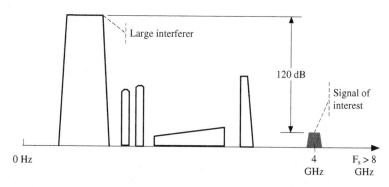

Fig. 2.3 Digitizing the signal of interest and adjacent bands

digitizer must have sufficient dynamic range to process both the strong and the
weak signals. An ideal digitizer provides about 6 dB of dynamic range per bit of
resolution. The digitizer would then have to provide well over 20 bits of resolution (e.g., 20 to resolve the interferer and more for the signal of interest).

- The digitizer must be very linear. Nonlinearity causes intermodulation between
  all the signals in the digitized band (see Fig. A-31 in Sect. A.9.6). Even a high
  order intermodulation component of a strong signal can swamp a much weaker
  signal.

- In the extreme example discussed so far (a 24 bit digitizer operating at
  120 GHz) real-time digital signal processing has to be applied to a data stream
  at $120 \times 10^9 \times 24 \approx 250$ GB/s. This is beyond the capabilities of modern
  processors and is likely to remain so in the foreseeable future.

Assuming all of these technical problems were solved,[4] the same radio could be
used to process any existing and expected future waveforms. However, it does not
mean that radio is optimal or suitable for a given application. The ideal SDR may
be perfect for a research laboratory, where physical size and power consumption
are not an issue, but completely inappropriate for a handheld device. The next few
chapters will deal with the implementation tradeoffs for different market segments.

Some of the earliest software-defined radios were not wireless. The soft modems used in the waning days of dial-up implemented sophisticated real-time signal
processing entirely in the software.

---

[4] Note that the A/D converter with these specifications violates the Heisenberg uncertainty
principle and is therefore not realizable. The maximum A/D precision at 120 GSPS is limited to
~ 14 bits [157] (see also footnote 6 of Chap. 10)

Chapter 3
Why SDR?

It takes time for a new technology to evolve from the lab to the field. Since SDR is relatively new, it is not yet clear where it can be applied. Some of the most significant advantages and applications are summarized below.

- *Interoperability.* An SDR can seamlessly communicate with multiple incompatible radios or act as a bridge between them. Interoperability was a primary reason for the US military's interest in, and funding of, SDR for the past 30 years. Different branches of the military and law enforcement use dozens of incompatible radios, hindering communication during joint operations. A single multi-channel and multi-standard SDR can act as a translator for all the different radios.
- *Efficient use of resources under varying conditions.* An SDR can adapt the waveform to maximize a key metric. For example, a low-power waveform can be selected if the radio is running low on battery. A high-throughput waveform can be selected to quickly download a file. By choosing the appropriate waveform for every scenario, the radios can provide a better user experience (e.g., last longer on a set of batteries).
- *Opportunistic frequency reuse (cognitive radio.)* An SDR can take advantage of underutilized spectrum. If the owner of the spectrum is not using it, an SDR can 'borrow' the spectrum until the owner comes back. This technique has the potential to dramatically increase the amount of available spectrum.
- *Reduced obsolescence (future-proofing).* An SDR can be upgraded in the field to support the latest communications standards. This capability is especially important to radios with long life cycles such as those in military and aerospace applications. For example, a new cellular standard can be rolled out by remotely loading new software into an SDR base station, saving the cost of new hardware and the installation labor.
- *Lower cost.* An SDR can be adapted for use in multiple markets and for multiple applications. Economies of scale come into play to reduce the cost of each device. For example, the same radio can be sold to cell phone and automobile manufacturers. Just as significantly, the cost of maintenance and training is reduced.

E. Grayver, *Implementing Software Defined Radio*,
DOI: 10.1007/978-1-4419-9332-8_3, © Springer Science+Business Media New York 2013

- *Research and development.* An SDR can be used to implement many different waveforms for real-time performance analysis. Large trade-space studies can be conducted much faster (and often with higher fidelity) than through simulations.

The rest of this chapter covers a few of these applications in more detail.

3.1 Adaptive Coding and Modulation

The figure of merit for an SDR strongly depends on the application. Some radios must minimize overall physical size, others must offer the highest possible reliability, while others must operate in a unique environment such as underwater.

For some, the goal is to minimize power consumption while maintaining the required throughput. The power consumption constraint may be due to limited energy available in a battery-powered device, or heat dissipation for space applications. The figure of merit for these radios is energy/bit [J/b].

Other radios are not power constrained (within reason, or within emitted power limits imposed by the FCC), but must transmit the most data in the available spectrum. In that case, the radio always transmits at the highest supported power. The figure of merit for these radios is bandwidth efficiency—number of bits transmitted per second for each Hz of bandwidth [bps/Hz]. Shannon's law tells us the absolute maximum throughput that can be achieved in a given bandwidth at a given SNR. This limit is known as the capacity of the channel (see Sect. A.3).

$$C(\mathrm{SNR}) = B \log_2(1 + \mathrm{SNR})$$

Shannon proved that there exists a waveform that achieves the capacity while maintaining an arbitrarily small BER. A fixed-function radio can conceivably achieve capacity at one and only one value of SNR.[1] In practice, a radio operates over a wide range of SNRs. Mobile radios experience large changes in SNR over short periods of time due to fading (see Sect. A.9.3). Fixed point-to-point radios experience changes in SNR over time due to weather. Even if environmental effects are neglected, multiple instances of the same radio are likely to be positioned at different distances, incurring different propagation losses. Classical radio design of 20 years ago was very conservative. Radios were designed to operate under the worst case conditions, and even then accepted some probability that the link would be lost (i.e., the SNR would drop below a minimum threshold). In other words, the radios operated well below capacity most of the time and failed entirely at other times.[2]

---

[1] Given the assumptions of fixed bandwidth and fixed transmit power.

[2] Some radios changed the data rate to allow operation over a wide range of channel conditions. Reducing the data rate while maintaining transmit power increases SNR. However, capacity was not achieved since not all available spectrum was utilized.

Adaptive coding and modulation[3] (ACM) was introduced even before the SDR concept (p. 19 in [5]). However, SDR made ACM practical. The basic idea follows straight from Shannon's law—select a combination of coding and modulation that achieves the highest throughput given the *current* channel conditions while meeting the BER requirement [6]. Implementation of ACM has three basic requirements:

1. Current channel conditions[4] must be known with reasonable accuracy.
2. Channel conditions must remain constant or change slowly relative to the adaptation rate.
3. The radio must support multiple waveforms that operate closer to capacity at different SNRs.

The first requirement can be met taking either an open-loop or closed-loop approach. In the open-loop approach, information about the channel comes from outside the radio. Examples include:

- Weather reports can be used to predict signal attenuation due to weather (e.g., if rain is in the forecast, a more robust waveform is selected).
- Known relative position of the transmitter and receiver can be used to predict path loss.

  - GPS location information can be used to estimate the distance to a fixed base station.
  - Orbit parameters and current time can be used to estimate the distance to a satellite

A closed-loop (feedback) approach is preferred if the receiver can send information back to the transmitter (e.g., SNR measurements). This approach is more robust and allows for much faster adaptation than open-loop methods. However, a bidirectional link is required, and some throughput is lost on the return link to accommodate the SNR updates. Almost all radios that support ACM are closed loop.

The second requirement is usually the greatest impediment to effective use of ACM. Consider a mobile radio using closed-loop feedback. The receiver estimates its SNR and sends that estimate back to the transmitter which must process the message and adjust its own transmission accordingly. This sequence takes time. During that time, the receiver may have moved to a different location and the channel may have changed. The problem is exacerbated for satellite communications, where the propagation delays can be of the order of ¼ second.[5]

---

[3] This technique falls in the category of Link Adaptation and is also known as dynamic coding and modulation (DCM), or adaptive modulation and coding (AMC).

[4] Channel conditions encompass a number of parameters (see Sect. A.9). The SNR is the only parameter used in the discussion below.

[5] A geostationary satellite orbits about 35,000 km above the earth. A signal traveling at the speed of light takes 3.5×10^7 m$/3 \times 10^8$ m/s $= 0.12$ s to reach it. Round trip delay is then approximately 0.25 s.

The channel for a mobile radio can be described by a combination of fast and slow fading. Fast fading is due to multipath (see Sect. A.9.3), while slow fading is due to shadowing. ACM is not suitable for combating fast fading. The rate at which SNR updates are provided to the transmitter should be slow enough to average out the effect of fast fading, but fast enough to track slow fading. The updates themselves can be provided at different levels of fidelity: instantaneous SNR at the time of update, average SNR between two successive updates, or a histogram showing the distribution of SNR between updates [7]. It is easy to show that employing ACM with outdated channel estimates is worse than not employing it at all, i.e., using the same waveform at all times (see next page).

In the following discussion we derive and compare the theoretical throughput for a link with and without ACM. For simplicity let us assume that the SDR supports an infinite set of waveforms, allowing it to achieve capacity for every SNR.[6] Consider a radio link where SNR varies from x to y (linear scale), with every value having the same probability. Let the bandwidth allocated to the link be 1 Hz. (Setting the bandwidth to 1 Hz allows us to use terms *throughput* and *bandwidth efficiency* interchangeably.)

Let us first compute the throughput for a fixed-function radio. This radio has to pick one SNR value, s, at which to optimize the link. When the link SNR is below s, the BER is too high and no data are received. When the link SNR is above s, the transmission is error-free. The average throughput of the link is then

$$P(\text{SNR} > s) \times C(s) + P(\text{SNR} < s) \times 0 = \frac{y - s}{y - x} \times \log_2(1 + s)$$

The optimal s, which maximizes throughput, is easily found numerically for any x, y. Numerical examples and graphs in this section are computed for $x = 1$ and $y = 10$ (0–10 dB). The capacity of the link varies between $\log_2(1 + x)$ (1 bps) and $\log_2(1 + y)$ (3.5 bps). For our numerical example, shown in Fig. 3.1, optimal s, $s_0 = 3.4$, and the average throughput is $\frac{10 - 3.4}{10 - 1} \times \log_2(1 + 3.4) = 1.6$ bps.

The average throughput is maximized but the system suffers from complete loss of communications for $(s_0 - x)/(y - x)$ percent of the time (over 25 %), which is not acceptable in many scenarios. For example, if the link is being used for a teleconference, a minimum throughput is required to maintain audio, while video can be allowed to drop out. In these scenarios a single-waveform radio must always use the most robust waveform and will achieve throughput of only $C(x)$ (1 bps).

Let us now compute the average throughput for an SDR that supports a wide range of capacity-approaching waveforms. Assuming the SNR is always perfectly known, the average throughput is simply

---

[6] This assumption is not unreasonable since existing commercial standards such as DVB-S2 define waveforms that operate close to capacity for a wide range of SNR (see Fig. 3.2).

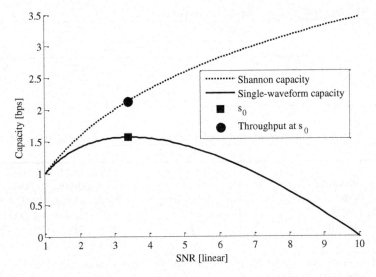

Fig. 3.1 Shannon and single-waveform capacity, s_0 is the optimal s. X-axis is the *target* SNR

$$E[C] = \frac{1}{y-x}\int_y^x C(s)ds = \frac{(1+y)\log_2(1+y) - (1+x)\log_2(1+x)}{y-x} - \frac{1}{\ln 2}$$

(equal to 2.6 bps for our numerical example).

ACM actually degrades the link performance if SNR estimates are not available, are out of date, or are inaccurate. Consider performance of ACM in a fast-fading channel when adaptation latency is longer than the channel coherence time. In that case, the waveform selection is uncorrelated to the channel conditions and can be considered to be random. Let the waveform selected by ACM be optimized for SNR $= a$ and let the actual SNR at the receiver equal b. No data can be sent when $a > b$, which happens on average half the time. The other half of the time, link throughput is given by $C(a)$. The average throughput is then $\frac{1}{2}E[C]$ (equal to 1.3 bps for our numerical example).

It is interesting to note that applying ACM with incorrect SNR updates results in slightly better throughput than always using the most robust waveform. Of course, always using the most robust waveform has the major advantage of guaranteeing no outages.

ACM is most effective if the expected[7] link SNR falls within the 'linear' region of the capacity curve (below 10 dB). The logarithmic curve flattens out at high SNR, meaning that capacity changes very little as SNR changes. If a radio operates in the 'flat' region of the capacity curve, then a single waveform is sufficient. Advanced techniques such as MIMO (see Sect. A.6) can be leveraged to increase

---

[7] ACM can be invaluable for the *unexpected* decrease in SNR.

Table 3.1 Numerical examples for ACM capacity

x [dB]	y [dB]	Single waveform		ACM	
		Guaranteed	Optimal	Optimal	Random
−5	5	0.4	0.8	1.4	0.7
0	10	1	1.6	2.6	1.3
10	20	3.5	3.9	5.6	2.8

Table 3.2 Commercial standards that employ adaptive coding (C) and modulation (M)[a]

Standard	M	C	Waveforms	Comments
DVB-S2	4	11	28	LDPC, QPSK—32 APSK
802.16e (WiMAX)	3	4	8	Turbo, QPSK, 16QAM, 64QAM
GPRS/EDGE	3	4	9	Convolutional, GMSK, 8PSK
802.11n (WiFi)	4	4	8	LDPC, {2,4,16,64}-QAM
1xEVDO-Rev B	3	5	14	Turbo, QPSK, 8PSK, 16QAM
HSDPA	3	4	6	Turbo, QPSK, 16QAM, 64QAM

[a] Number of waveforms is not equal to the product of modulations and code rates. Only a subset of all combinations is allowed.

throughput at relatively high SNR. Capacity for different SNR ranges and ACM scenarios is computed in Table 3.1.

Many commercial wireless standards have embraced ACM since it offers a significant improvement in average throughput at the cost of a modest increase in radio complexity. A few representative standards are summarized in Table 3.2.

The DVB-S2 standard was developed for satellite communications. This scenario is ideally suited for ACM since the channel SNR is expected to vary slowly,[8] allowing plenty of time for adaptation. The DVB-S2 standard defines 28 different waveforms which cover a 20 dB range of SNR. The bandwidth efficiency of these waveforms varies over almost an order of magnitude from 0.5 to 4.5 bps/Hz. As can be seen from Fig. 3.2, DVB-S2 waveforms achieve bandwidth efficiencies within 1–2 dB of capacity.

DVB-S2 has by far the largest number of waveforms of all current commercial standards. Interesting results reported in [8] indicate that more waveforms are not necessarily better[9] once real-world constraints such as non-ideal channel estimates

---

[8] A line-of-sight link is assumed. SNR changes are then due to either weather conditions or satellite orbit. Multipath propagation which causes fast fading is not a significant concern. Scintillation is assumed negligible.

[9] Some of the DVB-S2 waveforms are either redundant or suboptimal in which a waveform with higher bandwidth efficiency exists for the same SNR. These redundant waveforms do not mean that the standard was poorly designed. For example:

1. QPSK with rate 8/9 code requires 6.09 dB SNR and achieves 1.76 bps/Hz.
2. 8PSK with rate 3/5 requires 5.71 dB SNR and achieves 1.78 bps/Hz.
   Thus, the second waveform is always preferable to the first. However, some radios may not support 8PSK modulation and would therefore benefit from having QPSK at high code rates.

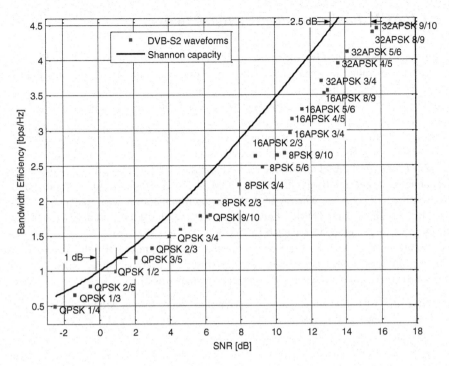

Fig. 3.2 Spectral efficiency of DVB-S2

Table 3.3 DVB-S2 waveforms with almost identical throughput

#	Modulation	Code rate	SNR [dB]	Throughput [bps/Hz]
1	QPSK	3/4	3.9	1.5
2	QPSK	4/5	4.5	1.6

and non-zero adaptation latency are taken into account. A link operating at the "hairy edge" of a given waveform becomes susceptible to even small fluctuations in SNR. Consider the waveforms in Table 3.3.

Only 0.6 dB separates these two waveforms, and the second one offers only 7 % higher throughput. If the SNR estimate is incorrect or drops by 0.6 dB, a block of data transmitted with waveform 2 will be lost. At least $1/0.07 \approx 15$ blocks have to be received correctly to compensate for the lost block and achieve the same average throughput as waveform 1. Larger spacing between SNR values required for different waveforms is acceptable, and in fact preferred, unless the channel is static and the SNR estimates are very accurate.

3.1.1 ACM Implementation Considerations

Although SDR is ideally suited for implementing ACM, a number of implementation issues must be considered. Adaptation rate is an important implementation driver. Fast ACM allows the code and modulation (CM) to change on a frame-by-frame basis. Slow ACM radio assumes that the CM changes infrequently and some time is available for the radio to reconfigure. Most modern wireless standards rely on fast ACM. Fast ACM is an obvious choice for packet-based standard such as WiMAX. DVB-S2 is a streaming-based standard that also allows CM changes on every frame. Fast ACM allows the radio to operate close to capacity even when the channel changes relatively fast. Fast ACM is also used for point to multi-point communications (i.e., when one radio is communicating simultaneously with multiple radios) using TDMA. Each of the links may have a different SNR, and the radio has to change CM for each time slot.

For both fast and slow ACM, a mechanism is required for the receiver to determine what CM was used by the transmitter for each set of received samples (frames or packets). Three most common mechanisms are:

1. Inserting the CM information (CMI) into each frame header. This is the mechanism used by DVB-S2.
2. Passing the CMI on a side channel (e.g., control channel). This is the mechanism used by WiMAX.
3. Passing the CMI as part of the data payload in some frames.

The first two approaches are suitable for fast ACM, while the third is used for slow ACM. Providing CMI for every frame is inefficient if the channel is known to change slowly. The overhead required to transmit CMI is then wasted most of the time. CMI must be received correctly with very high probability since an error in CMI always leads to complete loss of the frame it described. For example, in DVB-S2 CMI is encoded using a rate 7/64 code (45 bits are used to send 5 bits of CMI) and modulated with BPSK. This overhead is negligible[10] for the long frames used by DVB-S2, but could be unacceptable for shorter frames.

An all-software radio can easily support fast ACM since all of the functions required to implement each CM are always available. Supporting fast ACM on an FPGA-based SDR is more challenging. Consider three different FPGA reconfiguration options described in Sect. 6.2. Fast ACM can be supported if the FPGA contains all the functions required to implement each CM. If a different configuration has to be loaded to support a specific CM, the radio is not available during the reconfiguration time and fast ACM cannot be supported. Partial reconfiguration can be used to support fast ACM if the FPGA has enough resources to simultaneously support multiple (but not all) CMs.

---

[10] Worst case overhead occurs for the shortest frames. For DVB-S2, the shortest frame (16200 bit) transmitted using the highest modulation (32-APSK) requires a total of 3240 symbols. In that case the overhead is $45/3240 = 1\%$.

3.2 Dynamic Bandwidth and Resource Allocation

The ACM technique described above can be extended to vary the signal bandwidth. Channel capacity grows monotonically with the available bandwidth. Therefore, most single-user point-to-point radios use all the bandwidth allocated to them. However, if multiple users share the same spectrum, per-user bandwidth allocation becomes an important degree of freedom. Traditionally, frequency division multiple access (FDMA) approach was used to allocate the same amount of spectrum to each user. Modern radios allow each user to process different signal bandwidths at different times. There are three major techniques to achieve this:

1. A single-carrier radio can change its symbol rate, since occupied bandwidth is linearly related to the symbol rate. This approach was selected for the next generation of military satellites (TSAT) [9].
2. A radio may be allowed to transmit on multiple channels. For example, GSM defines each channel to be 150 kHz, but one radio can use multiple adjacent channels.
3. Orthogonal frequency division multiple access (OFDMA) is the technique used by many modern wireless standards to allocate subsets of OFDM (see Sect. A.4.7 and [10, 11]) subcarriers to different users. This is equivalent to (2) in the context of OFDM (i.e., subcarriers are orthogonal and separated by the symbol rate).

DBRA works well with ACM in a multi-user environment, where different users have uncorrelated SNR profiles. At any given time some users will experience low SNR, while others will have high SNR.

Consider a system in which all users are accessing real-time streaming data (e.g., voice or video). Each user must maintain constant throughput to avoid dropouts. The overall system is constrained by the total allocated bandwidth and total transmit power. High SNR users can adapt CM to achieve high bandwidth efficiency and can maintain the required throughput at a lower symbol rate. Lowering the symbol rate frees up bandwidth that can be allocated to the low SNR users as shown in Fig. 3.3. Alternatively, this goal could be achieved by allocating more power to the nominally low SNR user (thus increasing its SNR) and less power to the high SNR user.

The simple example below demonstrates the advantage of using DBRA and ACM instead of just power control. A more general treatment of the same problem, but in the context of OFDM, is presented in [12].

A radio is sending data to two users. The channel to user 2 has X times ($10 \log_{10} X$ dB) higher attenuation than the channel to user 1. Let the total power be $P = P_1 + P_2$. The total bandwidth is $B = B_1 + B_2$, which makes user 2's bandwidth $B_2 = B - B_1$. The first user's capacity is:

$$C_1 = B_1 \log_2 \left(1 + \frac{P_1}{N_0 B_1} \right)$$

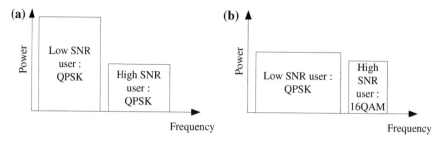

Fig. 3.3 Applying DBRA to a two-user channel **a** no DBRA, **b** with DBRA

Expressing the second user's capacity in terms of first user's bandwidth, we get

$$C_2 = (B - B_1) \log_2 \left(1 + \frac{P_2/X}{N_0(B - B_1)} \right)$$

And, since both users require the same throughput,

$$C_1 = C_2 = C$$

We want to find the bandwidth allocation that minimizes total power, P. Solving for P_1 and P_2, we get

$$P_1(B_1) = B_1 N_0 \left(2^{\left(\frac{C}{B_1} \right)} - 1 \right)$$

$$P_2(B_1) = (B - B_1)(N_0 X) \left(2^{\left(\frac{C}{B - B_1} \right)} - 1 \right)$$

Total power is then

$$P_{\text{DBRA}} = P_1(B_1) + P_2(B_1)$$

We can find $B_1 = B_{\text{opt}}$ that minimizes P numerically. Somewhat unexpectedly, B_{opt} is independent of N_0.

If only ACM is available and the symbol rate (i.e., bandwidth) for both users is the same, $B_1 = B_2 = B/2$, the second user has to be allocated X times more power to achieve the same capacity as the first user. Total power for the ACM-only case is then

$$P_{\text{NO DBRA}} = P_1 \left(\frac{B}{2} \right) + P_2 \left(\frac{B}{2} \right) = (1 + X) P_1 \left(\frac{B}{2} \right)$$

It is interesting to compare the power advantage, R, offered by DBRA

$$R = \frac{P_{\text{DBRA}}}{P_{\text{NO DBRA}}}$$

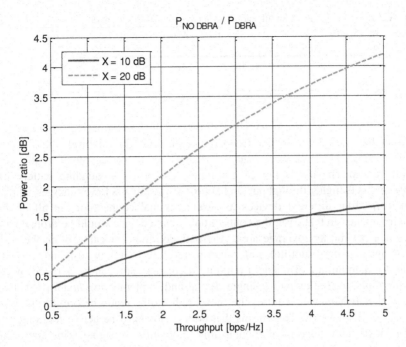

Fig. 3.4 Improvement offered by DBRA

Consider a numerical example, where $B = 1$ and $X = 10$ dB. A plot of R for a range of capacity values is shown in Fig. 3.4. As can be seen from this figure, DBRA offers only a modest improvement of about 1 dB, and only for high bandwidth efficiency. The improvement is more significant for a larger power difference among the users, $X = 20$ dB.

Another view of DBRA is offered in the numerical example below. For variety, let $B = 1$ MHz, $X = 20$ dB, and $C = 1$ Mbps. Solving for B_{opt}, we get the results in Table 3.4. For this scenario, $R = 1.1$ dB.

This somewhat disappointing result does not mean that DBRA is not a useful technique. For the TSAT program, DBRA was used to ensure that high-priority users can get high-throughput links even under very bad channel conditions (at the cost of lower or no throughput for other users).

3.3 Hierarchical Cellular Network

A mobile user has access to different wireless networks in different locations. For example, at home a user can access his wireless LAN; 4G cellular network is available outside his house; 2G network covers outside major metropolitan areas; and only satellite coverage is available in the wilderness. Each of these networks

Table 3.4 Optimal bandwidth allocation for a two-user DBRA scenario

User	Bandwidth [kHz]	Power [relative N_0]	SNR [dB]	Efficiency [bps/Hz]	Waveform
1	200	5.8	14.5	4.9	32-APSK, rate $= 9/10$
2	800	111	21.5	1.3	QPSK, rate $= 2/3$

uses a different waveform which is optimized for the target environment (e.g., 802.11 for LAN, LTE for 4G, GSM for 2G, Iridium for satellite) (Fig. 3.5).

The most 'local' network usually provides better service for those that cover wider areas. The user's device consumes less power (extending battery life), typically gets higher throughput, and almost always incurs lower charges. The user can carry four different devices to ensure optimal connectivity in all of these environments. In that case, the connection on each device would be dropped as he traveled outside its coverage area. Alternatively, the user can rely on the widest coverage network (satellite), suffer poor battery life, and high cost, and still not be able to maintain a call inside a building. Dropped calls can be avoided by integrating all four radios into a single device and implementing *handover* between them.[11] Soft handover requires the 'next' radio to be ready before the 'previous' radio is disconnected. This means that both radios must be active for some time. The 'next' radio must have enough time to acquire the signal. This approach is shown in Fig. 2.1.

An SDR combines all the separate radios into one, and almost surely simplifies handover between networks. The internal state (e.g., absolute time, or channel estimate if the two networks share the same carrier frequency) is preserved during handover, reducing the amount of time required to lock onto the new signal.

3.4 Cognitive Radio

The concepts of software-defined radio and cognitive radio (CR) were both proposed by Joseph Mitola III [13]. In this 'big picture' paper, Mitola envisioned radios that are not only reconfigurable but are also aware of the context in which they are being operated. Such radios consider all observable parameters to select the optimal set of communications parameters. Theoretically, one such radio could "see" a wall between the receiver and itself and choose to operate at a carrier frequency that would be least attenuated by that wall, while taking into account all other radios around it. The cognition cycle responsible for processing all inputs and making appropriate decisions is depicted in Fig. 3.6. The radio responds to external stimuli by first orienting itself to the urgency of the events. Some events (e.g., battery level critical) must be responded to immediately, while others

---

[11] Of course, the different networks have to support soft handover at the protocol layer.

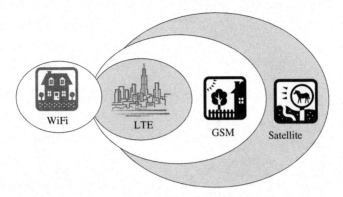

Fig. 3.5 Hierarchical network coverage

(e.g., motion detected) allow the radio some time to come up with an optimal solution. For example, the ACM technique described previously uses only the inner (Observe, Orient, Act) loop. A key part of a true cognitive radio is ability to learn (in the Learn state) what the outcome of a given decision will be. For example, it could automatically learn which CM is optimal for a given SNR. The decisions are then driven by all previously learned data and a set of predefined constraints which depend on the radio's context (e.g., carrier frequency 1.5–1.6 GHz can be used if the radio is in Europe, but 2.0–2.1 GHz if the radio is in the US). The combination of predefined and learned data allows a radio to work out-of-the-box and provide better performance over time.

We are still a long way from implementing such an intelligent radio, and it is not at all clear whether there is a need for it. For now, the ambitions for CR have been scaled back to a "spectrum aware" radio [14]. Design of a spectrum-aware radio is motivated by the realization that spectrum is a precious shared resource. Most of the desirable spectrum (see Sect. A.2) is allocated to users (by the FCC in the United States [15] and equivalent organizations around the world). For example, a local FM radio station is allocated a band at 98.3 MHz within a certain geographic area. As new technologies such as cellular phones become available, they compete for spectrum resources with incumbent (also known as primary) users. The latest smartphones can process data rates of many Mbps to provide users with streaming video. A significant amount of spectrum is required to transmit that much data. Bands allocated to cell phones were sized with only voice traffic in mind and are heavily congested. Police, emergency medical, and other public safety personnel also need more spectrum than their existing allocation, especially during a large-scale emergency [16].

Spectrum management agencies have long known that allocated spectrum is underutilized. At any given time, only a small fraction of the allocated bands are in use by the incumbents. In many areas, not all TV and radio stations are active. Some bands (e.g., those allocated for emergency satellite control) are used only a few minutes a day. A spectrum-aware CR is allowed to transmit in the allocated

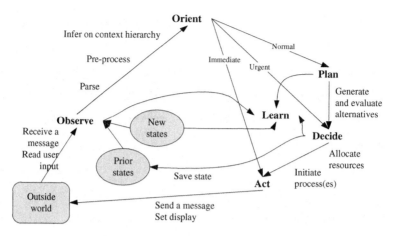

Fig. 3.6 Cognition cycle as envisioned in [13]

band, but must vacate that band if an incumbent user appears. Note that the incumbents are not required to notify other radios of their intent to transmit. The onus is on the cognitive radio to sense the incumbents.

The problem of spectrum sensing is a very active research area [17]. Sensing can be accomplished by a single radio, or multiple CRs can each sense and share their results. When multiple CRs attempt to use the same spectrum, they must sense not only the incumbent, but each other as well. Once a radio determines that a band is free, it must configure its transmitter to the correct carrier frequency and bandwidth. The transmission must not be so long that the CR fails to detect a newly active incumbent. The duration of the sensing and transmitting phases of the cycle depend on both technical and regulatory constraints. For example, if the FCC regulations require a cognitive radio to vacate the band within 100 ms after the incumbent appears, the cycle must be no more than 100 ms. The amount of time required to reliably sense the incumbent depends on the sensing algorithm, knowledge of the incumbent signal, and the hardware implementation.

Perhaps the biggest challenge in spectrum sensing is known as the "hidden node" problem. It can be caused by many factors including multipath fading or shadowing experienced by CRs while sensing incumbents. Figure 3.7 illustrates the hidden node problem, where the dashed circles show the operating ranges of the incumbent and the CRs. Here, the CR causes unwanted interference to the victim user because the primary user's signal was not detected by the CR.

A generic CR consists of a number of subsystems as shown in Fig. 3.8. A cognition engine (CE) is responsible for taking all available information and finding an (optimal) solution. If the requirements cannot be met (i.e., no solution exists), the radio should remain inactive. The SDR serves as the physical radio platform that will process the waveforms chosen by the CE. Inputs to the CE may include:

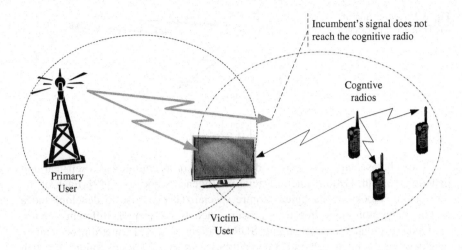

Fig. 3.7 The 'hidden node' problem for spectrum sensing

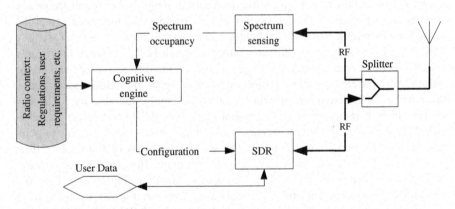

Fig. 3.8 Block diagram of a cognitive radio

- FCC regulations (e.g., vacate in 100 ms, maximum transmit power is 1 W, etc.)
- User bandwidth requirement (e.g., need 1 Mbps to stream video).
- Remaining battery level and how long the battery should last. If the transmission given available bandwidth would require too much power, the radio should not transmit.
- Spectrum-sensing output that includes a list of bands open for transmission.
- Geographical position. Position could determine the applicable regulations (e.g., different in Europe and USA).
- Radio environment could be derived from the geographical position. The optimal modulation in an urban setting is different from that in a rural setting.

Fig. 3.9 Application of
MUD to spectrum sharing

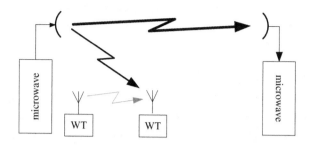

CE can be as simple as a hard-coded flowchart or as complex as a self-learning genetic algorithm [18] or other artificial intelligence algorithms [19]. Each of the three major blocks in the figure require specialized expertise to develop, and a system integrator may choose to procure each block from a different vendor. A standard method for exchanging SDR configuration and spectrum occupancy results would facilitate subsystem compatibility among different vendors. No such standard has yet been accepted by the SDR community, but one proposal is discussed in Sect. 7.3.

As spectrum becomes more crowded, simply moving to a different frequency becomes too restrictive. The latest work in CR considers simultaneous use of the same band by multiple users. A technique known as multi-user detection (MUD[12]) can be applied to separate multiple overlapping signals [20]. The separation is only feasible under certain conditions (e.g., signals from different users should be substantially different to allow projection of each signal onto independent basis). An advanced CE determines whether a particular band is suitable for MUD and ensures that all radios using the band can implement MUD or tolerate the additional interference. Consider a scenario with a high-power one-way point-to-point microwave link coexisting with low-power walkie-talkies (WT). Interference from the walkie-talkies to the microwave link is negligible. However, the microwave link easily overpowers the walkie-talkies, as shown in Figure 3.9. The WT waveform is very different from the microwave waveform. Further, the microwave waveform is received by the WT with high power and high SNR. The WT can then apply MUD to suppress the interfering microwave signal. The amount of suppression depends on the MUD algorithm, channel conditions, etc. These algorithms are computationally intensive and increase the power consumption of the DSP subsystem in WT. However, the reduction in WT transmit power made possible by the interference mitigation often more than makes up for the power cost of the mitigation [21].

---

[12] MUD can be extremely computationally expensive and has not been implemented in handheld devices yet. However, it is only a matter of time....

3.5 Green Radio

Radio engineers usually focus on optimizing a single communications link. For example, power consumption at a user terminal is optimized to increase battery life. A new concept called *green radio* addresses the power consumption of the entire communications infrastructure. For the case of cellular telephony that includes power used by all the base stations, handsets, and network equipment. The network operators benefit from reduced power costs and positive publicity.

The radio interface accounts for about 70 % of the total power consumption and therefore offers the largest potential for power savings. SDR provides the controls to adapt the radio interface to minimize total power. A cognitive radio approach described above can be applied to tune the controls [22]. The fundamental tradeoffs enabled by SDR are described in [23] and shown in Figure 3.10. The metrics identified in that paper are:

1. Deployment efficiency (DE) is a measure of aggregate network throughput per unit of deployment cost. The deployment cost consists of both the hardware (e.g., base station equipment, site installation, backhaul) and operational expenses (e.g., power costs, maintenance). Many closely spaced base stations allow each to transmit at a lower power level. Received power decreases with (at least) the square of the distance and distance decreases linearly with number of base stations. Thus, total power can be reduced by increasing the number of base stations, although the deployment cost increases.

2. Spectrum efficiency (SE) is a well-understood metric (see Sect. A.2). Energy efficiency (EE) is related to SE using Shannon's capacity equation, $(EE) = \frac{SE}{(2^{SE}-1)N_0}$. This is a monotonically decreasing function. It appears that the most robust waveforms (lowest SE) result in the highest EE. However, if one considers the static power consumption of the transceiver electronics, the EE at very low SE actually decreases. It is easy to see that if one waveform takes 1 min and another takes 1 s to transmit the same amount of data, the radio using the first waveform must remain powered on for much longer. Thus, there exists an optimum value of SE that maximizes EE.

3. Bandwidth (BW) and power (P) are two key constraints in a wireless system. Using Shannon's capacity equation, we can show that $P = BW \times N_0 2^{\frac{R}{BW}-1}$, which is a monotonically decreasing function. Thus, it is always beneficial to use all available bandwidth to reduce required power. For example, if a base station is servicing just a few users, each user can be allocated more bandwidth [21]. The spectrum-aware radio approach described in the previous section can be used to increase the amount of available bandwidth (also see [24]). The DBRA approach described in Sect. 3.2 can then be applied to allocate bandwidth to users. Practically, this tradeoff is somewhat more complicated because in practical systems power consumption increases when processing a larger bandwidth (i.e., a 1 Gsps ADC consumes much more power than a 1 Msps ADC).

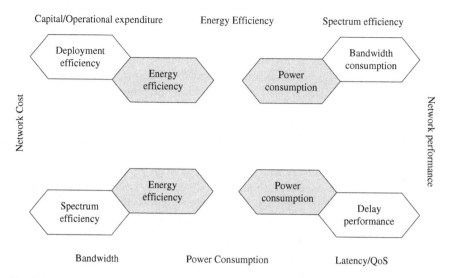

Fig. 3.10 Fundamental tradeoffs for green radio [23] (*shaded blocks* indicate negative effect of improving the *white blocks*)

4. Delay (DL) is the last metric that is considered in optimization of energy efficiency. In the simplest case, delay can be related to throughput, which is in turn related to spectrum efficiency. Tolerating larger delays (slower transmission) allows reduction in power. The effect is even more significant if we consider operation in a fading channel. When a user is experiencing a fade, it requires more power to deliver data. If larger delays can be tolerated, the system can schedule transmissions to users with good channel conditions first. This approach, closely related to opportunistic beamforming [25], can significantly increase capacity and reduce power at the same time.

These tradeoffs are enabled by the degrees of freedom provided by a SDR. The more degrees of freedom, (also known as *knobs* in CR literature), the closer the system can achieve optimal power consumption [26]. The basis for power consumption optimization at the physical layer is adaptive coding and modulation (see Sect. 3.1). The work in [26] delves deeper into the details of a waveform to study second-order parameters such as pilot symbols and number of antennas.

3.6 When Things go Really Wrong

True SDR (as opposed to software-controlled radio) really excels when unexpected problems occur and the radio cannot be replaced or repaired. The best examples come from radios used in space, where the cost of failure is very high and no options for hardware repair exist.

3.6.1 Unexpected Channel Conditions

The Cassini-Huygens mission was launched in 1997 to study Saturn and its moons. The Huygens probe was designed to separate from the main spacecraft and land on Saturn's moon, Titan, 8 years after the launch. Huygens would transmit data to Cassini, which in turn would retransmit it to the Earth. Long after launch, engineers discovered that the communication equipment on Cassini had a potentially fatal design flaw, which would have caused the loss of all data transmitted by Huygens.[13] The radio on Cassini could not correctly compensate for the high Doppler shift experienced by Huygens during the rapid descent to the surface [27, 28]. A small change to the symbol tracking algorithm in the Cassini radio could have fixed the problem, but the radio was not an SDR and could not be changed. Luckily, a solution was found by changing the flight trajectory of Cassini to reduce Doppler shift and all of the science data was returned. As noted in [28], "Spacecraft systems need to have an appropriate level of reconfigurability in flight. This anomaly would have been easy to solve if there has been even a modest amount of reconfigurability in the [Cassini radio]". An SDR would have met this modest requirement and could have allowed even more science data return by using ACM during the descent.

3.6.2 Hardware Failure

Most software-controlled radios support a set of waveforms from a single family (e.g., linear modulations such as BPSK—16QAM or nonlinear modulations such as GMSK, but not both). Very few scenarios require a radio to switch to a different waveform family. One illustrative example that did require changing waveform family involves a transmitter with a faulty carrier synthesizer. An ideal synthesizer generates a pure sinusoid at the desired frequency, or equivalently the phase of the output is perfectly linear. A component failure in a synthesizer caused the output phase to vary randomly at a very high rate. The phase tracking loop in the receiver could not keep up with the rapid changes, making demodulation impossible. Unlike the previous example (Cassini), changing the loop parameters or algorithm would not be sufficient. Analysis showed that coherent demodulation was impossible without significantly increasing transmit power (which was not an option). A waveform that can be demodulated noncoherently (see Sect. A.4.5) does not require a phase tracking loop and could solve the problem. A waveform such as differential BPSK (DBPSK) requires almost no additional DSP resources compared to BPSK and is easy to implement on the existing hardware. However,

---

[13] Doppler shift results in both a frequency shift (compensated by Cassini radio) *and* symbol rate shift (not compensated). The symbol rate shift is negligible for small Doppler shifts and was overlooked by the engineers.

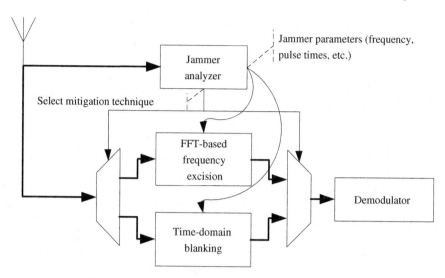

Fig. 3.11 Cognitive antijam technique implemented in an SDR

even DBPSK suffered significant degradation is due to the rapid phase variations. FSK modulation can also be detected noncoherently and was found to be a better solution to the problem.

3.6.3 Unexpected Interference

Once a radio is out in the field, it can experience intentional or unintentional interference (also see A.9.5). A rich set of interference mitigation techniques is available. However, each technique is only applicable to a specific type of interference. SDR provides radio engineers with powerful options for interference mitigation. Consider a fielded military radio that experiences a previously unknown interferer (a jammer). The radio can characterize the jammer waveform or just capture a sequence of samples when the jammer is present. The information is then passed to an analyst who can select the appropriate existing technique or develop a new one. The radio is then updated with software to mitigate the jammer. An autonomous jammer analyzer and mitigation technique selection approach, called *cognitive antijam*, is described in [29, 30]. The analyzer can distinguish between wideband burst and narrowband continuous jammers, as shown in Figure 3.11.

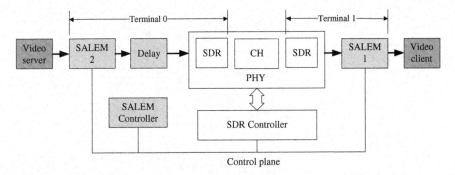

Fig. 3.12 Architecture of integrated test bed

3.7 ACM Case Study

The physical layer techniques discussed above are only a part of a complete SDR. This section describes a case study aimed at evaluating the effectiveness of a combination of techniques in a complete system [31]. Three mitigation techniques were considered singly and together:

1. Adaptive coding and modulation (ACM) at the physical layer.
2. Automatic repeat request (ARQ) at the network layer.
3. Automatic codec adaptation (ACA) at the application layer.

The system in question is a next-generation military satellite used to provide real-time video feeds to commanders in the field. As discussed in Sect. A.9, most wireless links must contend with channel fades and limited bandwidth. Satellite links have some unique properties (e.g., large propagation delays) that distinguish them from terrestrial links.

A variety of techniques (e.g., FEC, channel interleaving, and ARQ) have been proposed to mitigate satellite channel impairments, but it is not clear what effect these mechanisms will have on upper layer protocols and end-user applications. It is crucial to evaluate the cross-layer performance of these mitigation techniques before they are deployed. A test bed consisting of a software-defined radio, a real-time wireless channel emulator (CH), and a network-layer emulator (SALEM) was developed to study end-to-end system performance. The SALEM also implements network-layer mitigation techniques such as ARQ. Figure 3.12 shows the architecture of the integrated test bed for emulating future military communication scenarios. The test bed and results are described in detail below.

Commercial test equipment that generates and analyzes packets (Agilent N2X [32]) was used to quantify the performance of cross-layer mitigation techniques. The techniques were compared based on standard network-layer metrics such as packet loss ratio (PLR), latency, and throughput. Real-time performance of combinations of multiple mitigation schemes was observed with a streaming video application.

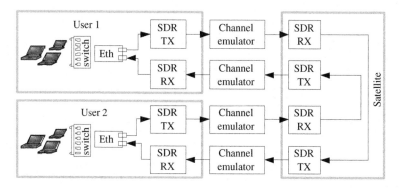

Fig. 3.13 Physical layer emulation for the test bed

3.7.1 Radio and Link Emulation

The physical layer emulator (PHY) consists of SDR and CH as shown in Fig. 3.13. It provides on-the-fly configurability to facilitate performance assessment under various communication scenarios. Four SDR transceivers are used to model a full duplex link between two users over a fully processed satellite payload. User SDRs provide a standard Ethernet interface. Each of the four links can use a different waveform. A wide range of waveforms (BPSK—64QAM) and data rates (1 kbps— 40 Mbps) is supported. Additionally, forward error correction (e.g., Reed-Solomon, Viterbi, Turbo, LDPC, etc.), interleaving, and encryption are all supported [33]. The SDR transceiver is configured using an XML description of a waveform (described in Sect. 7.3). Waveforms can be reconfigured in a few milliseconds [33].

The *channel emulator* applies realistic impairments to the transmitted waveform in real time. Both flat and frequency-selective (multipath) fading models are supported. The channel emulator is implemented using a combination of software and hardware: channel coefficients are computed in software and applied to the waveform in hardware (Fig. 3.14). The software can generate a range of channel scenarios such as line-of-sight, urban environment, and nuclear scintillation, and also allows models to be imported via precomputed data files.

SALEM provides link-layer capabilities, such as link-layer error mitigation (ARQ), adaptive modulation and coding (ACM), application codec control (ACC), and packet segmentation/reassembly [34]. Figure 3.15 depicts the architecture of SALEM integrated with PHY. In the integrated test bed, a delay node is inserted between SALEM2 and PHY to emulate the propagation delay incurred at the satellite up and down links (0.25 s).

ARQ is a link-layer error mitigation technique based upon packet retransmission. SALEM1 transmits an acknowledgment (ACK) frame for every packet it receives from SALEM2.[14] If SALEM2 does not receive an ACK frame for a

---

[14] For clarity, only the link from user 1 to user 2 is described.

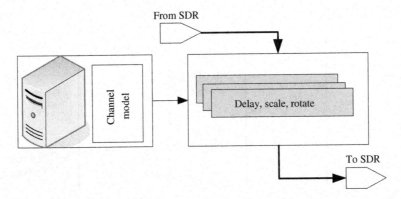

Fig. 3.14 Channel emulator

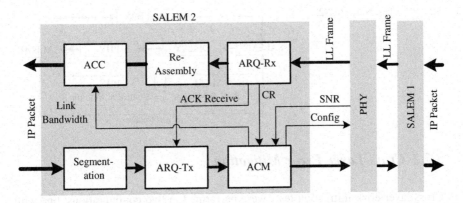

Fig. 3.15 SALEM block diagram

predetermined interval, it retransmits the unacknowledged frame. Retransmission is attempted up to a predetermined number of times.

SDR RX constantly monitors the SNR and periodically reports it to the ACM controller in SALEM1. SALEM1 then creates and sends a channel report (CR) to SALEM2. SALEM2 searches a database to find the optimal waveform supported for the SNR. If the optimal waveform is different from the current one, SDR TX is reconfigured. If a change in CM results in a change in the available bandwidth, SALEM2 sends a message to the video server to change the codec to match the video data rate to available bandwidth.

An open-source media player (VLC) was used for video encoding and play-back. A metric called the video quality indicator (VQI) was implemented in VLC to assess the quality of a video stream. A codec control interface was also implemented in VLC to dynamically change the video compression per control messages from SALEM. Message exchange for ACM with video compression adaptation is shown in Fig. 3.16.

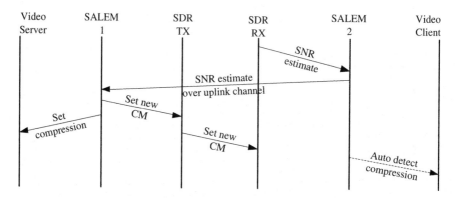

Fig. 3.16 Message passing for ACM and video compression adaptation (time flows down)

Fig. 3.17 Test bed with N2X

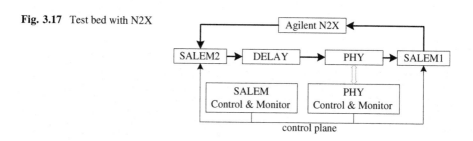

3.7.2 Cross-Layer Error Mitigation

Cross-layer error mitigation tests were performed in two configurations: one with Agilent N2X^{TM} and the other with a real-time streaming video application. The following subsections will describe in detail the test scenarios, including the channel model, various physical-, link-, and application-layer parameters, and corresponding results.

3.7.2.1 Performance Evaluation with a Traffic Analyzer

In this test, N2X is both the packet source and sink for the testbed, enabling automated measurement of PLR, latency, and throughput (RxRate) (Fig. 3.17). Two types of fading channel were modeled, short fades and long fades, as listed in Table 3.5. In each model, the channel SNR follows a saw-tooth trajectory (linear in dB) between the minimum and maximum values over some duration. FEC was disabled for this test and modulation was allowed to vary between BPSK and 16QAM when ACM was enabled and was fixed to 16QAM when ACM was disabled.

Table 3.5 Fading channel models

Model	Max SNR (dB)	Min SNR (dB)	Duration (s)
Short fade	35	17	5
Long fade	35	17	30

Table 3.6 Test results with N2X

Case ID	Channel model	ACM	ARQ	PLR (%)	RxRate (Mbps)	Avg. Lat. (ms)	Max. Lat. (ms)
1.1	Long	Off	Off	14.07	0.83	252	253
1.2	Long	On	Off	0	0.99	253	255
1.3	Long	Off	On	7.05	0.91	951	5568
1.4	Long	On	On	0	1.0	328	1317
2.1	Short	Off	Off	11.27	0.84	252	253
2.2	Short	On	Off	8.83	0.82	253	255
2.3	Short	Off	On	0	0.98	1528	3506
2.4	Short	On	On	0	0.94	1188	3144

Table 3.6 presents the test results with N2X, with a nominal transmit rate of 1 Mbps. The ARQ retransmission interval was set to 550 ms, and the maximum retransmission count was 7.

In case 1.1, about 14 % of packets did not survive the fading channel while the SNR was low, which resulted in a lower Rx Rate. Average and maximum latencies include both propagation and processing delays.

In case 1.2, the modulation automatically adapted to the current SNR, which significantly reduced the packet losses. In case 1.3, ACM was disabled while ARQ was enabled. Despite retransmissions by ARQ, about 7 % of packets were lost. This demonstrates that during long fades, even 7 retransmissions were not enough to save 7 % of packets. The average latency of 951 ms indicates each packet is retransmitted about 1.3 times on average.[15] The maximum latency reflects the maximum number of retransmissions plus processing time and queuing delays. Compared to case 1.1, ARQ reduced the PLR by about 50 % at the cost of increased latency. In case 1.4, both ACM and ARQ were enabled. Most of the packets made it through the fading channel, and packets dropped during the mode change were recovered by ARQ, which resulted in zero PLR. The increases in the average and maximum latencies compared to those from cases 1.1 and 1.2 were due to retransmission of lost packets during the mode change.

In cases 2.1 through 2.4, the channel fluctuated much faster than in the previous cases. Case 2.1 lost about 11 % of packets, which is comparable to case 1.1. In case 2.2, about 9 % of packets were lost despite ACM. Case 2.2 corresponds to the 'random SNR' scenario discussed on page 13.

---

[15] Propagation delay is 250 ms and the retransmission interval is 550 ms. The average number of retransmissions can be estimated as $(951-250)/550 = 1.3$.

Table 3.7 Test results with real-time streaming video

ID	ACM	ARQ	ACA	SNR [dB]	Mod	Data rate [Kbps]	Frame rate [FPS]	Frame size [B]	VQI
3.1	Off	Off	Off	36	16-QAM	873	31.2	2661	100
3.2	Off	Off	Off	18	16-QAM	344	10.0	2126	45.4
3.3	On	Off	Off	18	QPSK	580	19.8	2144	51.9
3.4	On	Off	On	18	QPSK	579	29.9	1349	89.4
3.5	On	On	On	18	QPSK	628	33.6	1448	99.8

In both cases 2.3 and 2.4, ARQ effectively recovered all of the lost packets caused by channel fades. ACM slightly reduced latency in case 2.4. The Rx Rate in case 2.4 (0.94 Mbps) turned out to be less than expected and thus needs further investigation.

3.7.2.2 Performance Evaluation with Real-Time Video Streaming

Cross-layer performance of ACM, ARQ, and ACA techniques was evaluated using a real-time streaming video application. A VLC application running on the video server transmitted an MPEG4-encoded video stream at the rate of 875 kbps to another VLC application running on the video client (see Fig. 3.12). The video client displays the received video and gathers MPEG frame statistics. Note that in this emulation, the SNR was manually lowered to emulate a channel fade.

Table 3.7 shows the test results for various cases. Each case was defined by a combination of ACM, ARQ, and ACA options and an SNR value. The data rate in the table indicates that measured at the server. In the following discussion, a frame refers to an MPEG frame, not a link-layer frame.

Initially, SNR was set to 36 dB and the modulation was set to 16 QAM (case 3.1). No frames were lost in this best-case scenario. In case 3.2, the SNR was set to 18 dB and ACM was not enabled. The modulation remained at 16 QAM, which caused more than 50 % of the packets to be lost, resulting in data rate of 344 kbps and frame rate of 10 fps. The VQI dropped to 45.4 and the video was garbled. ACM was enabled in case 3.3 and the modulation automatically switched to QPSK. In this case, since QPSK is more robust, more packets survived, resulting in the data rate of 580 kbps and the frame rate of 19.8 fps. However, now the bandwidth of the channel was reduced by 50 % and thus a lot of packets were dropped, resulting in video quality that was still very poor as indicated by a VQI of 51. Configuring the video codec to reduce the data rate by half (case 3.3, note that the average frame size is halved) improved VQI to 89. Some packets were still lost due to imperfect channel conditions. ARQ was enabled in case 3.5. ARQ retransmissions caused the data rate to increase (628 vs. 579 kbps). Average frame size indicates that the application codec generates the same streaming rate. The VQI improved to almost 100 and the video was excellent.

Video quality in case 3.1 was still noticeably better than in case 3.5 even though their VQI values are almost the same. This is because video was generated at a lower resolution to achieve a lower data rate, which is reflected in the smaller average frame size in case 3.5.

These experiments demonstrated the benefit of multiple, cooperating cross-layer adaptive techniques to mitigate channel effects.

Chapter 4
Disadvantages of SDR

The advantages of SDR are described throughout this book. The question then is: "Why isn't every radio already a SDR?" No technology is without its disadvantages and challenges, and some of them will be discussed in this chapter.

4.1 Cost and Power

The most common argument against SDR is cost. The argument is particularly important for high-volume, low-margin consumer products. Consider a garage or car door remote opener key fob. This extremely simple device has one and only function. The mechanical design—a single button—precludes the addition of new functionality in the future. Millions of essentially identical devices are sold every year, amortizing the development cost of an ASIC. The cost of the ASIC is then proportional to the size of the chip, which is in turn a function of the chip complexity. An SDR chip for garage door openers could be used in many other devices as well, but the increased market volume would not drive the cost of the chip down. In fact, since an SDR is necessarily more complex than a single-function radio, the SDR chip cost would be higher. The same cost argument has been made for AM and FM radio receivers. These products are also high-volume and potentially very low cost.[1]

---

[1] However, AM/FM receivers are often integrated into larger systems (e.g., car audio systems, personal audio players) which require a significant amount of signal processing. If the signal processing is already available, the incremental cost of making AM/FM receiver software-based is small. Moreover, unlike garage door openers, new waveforms are becoming available in the AM/FM channels. New digital radio standards such as HD radio [264] in the US and DAB [265] in Europe provide higher audio quality and new features while coexisting with legacy AM/FM. Unfortunately, since SDR was not used for older receivers, they cannot process new signals. While incompatibility with new waveforms is irrelevant for a one-dollar radio, it is quite inconvenient for a $400 car stereo.

E. Grayver, *Implementing Software Defined Radio*,
DOI: 10.1007/978-1-4419-9332-8_4, © Springer Science+Business Media New York 2013

The second most common argument against SDR is increased power consumption. Two sources contribute to higher power consumption in an SDR: increased DSP complexity and higher mixed-signal/RF bandwidth. Power consumption in an FPGA or GPP used to implement flexible signal processing is easily 10 times higher than in an equivalent ASIC.[2] Wideband ADCs, DACs, and RF front ends required for SDR consume more power than their narrowband equivalents. The difference is especially dramatic for wideband power amplifiers, which account for at least 70 % of the total power in a radio. A wideband amplifier is often half as efficient as a narrowband one.[3] Power consumption in the ADCs varies approximately linearly with bandwidth, but is less than linear for DACs. A wideband ADC often also requires higher dynamic range (more bits) to tolerate in-band interferers.[4] Increasing ADC dynamic range significantly increases the power consumption [35]. However, waveform adaptation enabled by SDR can reduce the required transmit power; thereby saving power (see Sect. 3.5).

Cost and power consumption arguments are combined when considering the amount of signal processing margin to be included in an SDR. The margin is defined as the ratio of the DSP horsepower (e.g. measured in FLOPS) provided by an SDR relative to that required for the baseline set of waveforms. For example, an SDR is initially developed to support two waveforms, with the more complex one requiring 1 GFLOPs. A 100 % margin requires the SDR to provide 2 GFLOPs. It is difficult to predict what waveforms a user may wish to host on an SDR and even a small underestimate of the margin can preclude the 'must have' future waveform. Consider some of the early deep space missions that relied on simple and computationally non-demanding FEC codecs. Thirty years ago the designers could not imagine the enormous computational requirements of modern iterative codecs. It would have been (then) unreasonable to include enough DSP margin to enable an upgrade to the powerful codecs of today. Lack of sufficient DSP margin is a very strong argument against the 'future-proof' promise offered by SDR.

4.2 Complexity

One generic argument against SDR is the additional complexity it requires. The complexity argument has at least three components:

- *Increased time and cost to implement the radio*. It takes more engineering effort to develop software and firmware to support multiple waveforms than to support

---

[2] As transistors continue getting smaller, DSP power consumption is becoming an ever smaller part of the total radio power consumption.

[3] Development of adaptive impedance matching networks based on MEMS components [267] or varactors [266] enables wideband *tunable* power amplifiers. These amplifiers provide a relatively narrow instantaneous bandwidth which can be tuned over a wide range (also see Sect. 10.4).

[4] Alternatively, a tunable RF filter is required to remove the interferers. Such filters are not yet available.

just one. Some argue that the increase in complexity is super linear (i.e., it takes more than twice as long to implement a radio that supports two waveforms than to implement two radios that each support one waveform). This claim is not unreasonable if the radio has to conform to a complex standard such as JTRS (see Sect. 7.1). Specialized expertise required to develop on a particular SDR platform may disappear soon after the radio is delivered to the customers. Developing new waveforms for that platform in the future can easily be more expensive than starting from scratch.[5]

- *Longer and more costly specifications and requirements definition.* An SDR design has to support a set of baseline waveforms but also *anticipate* additional waveforms. Some DSP resource margin must be provided to support future waveforms.
- *Increased risk.* At least two sources of risk must be considered:

  – Inability to complete the design on-time and on-budget due to the concerns presented above. Since SDR is still a relatively new technology, it is more difficult to anticipate schedule problems.
  – Inability to thoroughly test the radio in all of the supported and anticipated modes.

Testing SDR is a very active area of research. In the author's opinion, it is the single strongest argument against SDR. How does one test a radio that supports a huge number of waveforms? It is clearly not possible to test every combination of supported parameters (modulations, codes, data rates, etc.). Defining 'corner' cases is nontrivial since interaction between components is not obvious (e.g., is the performance of a tracking loop stressed at low rates or high rates?). Combinations of FEC block lengths and time constants of adaptive loops introduce a new level of complexity.[6] Errors in SDR design can affect not only the faulty radio, but also the entire network. In fact, a wideband SDR could conceivably transmit a signal in an unauthorized band, thereby jamming other life-critical signals.[7] Testing an SDR in steady-state (i.e., once a waveform has been selected) is not sufficient. Problems may be observed only when switching from one specific waveform to another.

Concerns with security and information assurance problems in an SDR are related to the complexity argument. Most modern wireless standards include some form of encryption and authentication. We expect our phone conversations to be relatively immune from interception. The ability to change many aspects of a waveform also implies that the security-related aspects could be changed. It is not

---

[5] The author is aware of at least one system that relied on software tools no longer available today. Even if the old tools could be procured and made to run on modern computers, the engineers with expertise in these tools are hard to find. Clearly, this problem is only a concern for SDRs with long life cycles such as those on satellite payloads.

[6] The author is aware of one *fielded* system that exhibited poor performance when the memory of a FEC decoder happened to be close to the period of phase noise from the front end.

[7] A bug in the spectrum-sensing block or in the radio constraints database for a cognitive radio (see Sect. 3.4) can make the radio transmit in the GPS band.

difficult to imagine a virus that modifies the SDR configuration to disable encryption, or even to have it transmit a copy of the signal at another frequency. These intrusions would be difficult to detect since the user experience is not affected. The problem is even more important for military radios that handle classified information. A virus that disables[8] all SDRs during a military operation would be a potent weapon indeed. However, there is nothing special about SDR security and well-understood techniques can be applied to the download of new waveforms into a radio [36]. The same infrastructure that is currently used for over-the-air distribution of secret keys can be applied to SDR security.

4.3 Limited Scope

Finally, it is important to keep in mind that SDR only addresses the physical layer. The user cannot take advantage of link throughput improvements made possible by SDR without cooperation from upper layers. One example demonstrating the need for cross-layer adaptation is described in Sect. 3.7.2.2. Another example is described below.

The ubiquitous network protocol, TCP/IP, was developed for fixed-rate channels. A TCP transmitter sends data packets at the highest possible rate. The receiver responds with an acknowledgment packet (ACK) as it receives the data. If the transmitter does not get an acknowledgement, the protocol determines that the link is congested and quickly decreases the data rate. If more acknowledgement packets are not received, transmitter rate falls even further. The rate increases slowly once the ACK packets are received once again. Performance of TCP in the context of time-varying throughput caused by SDR adaptation to the environment has been extensively studied. Results show that end-to-end throughput is much lower than the physical layer throughput. Moreover, some combinations of channel fade rate and TCP settings lead to instability and throughput drops dramatically. Achieving the *average* throughput made possible by ACM may also require large buffers at the network routers or in the SDR.

Consider a *fixed-rate* data source[9] with the rate equal to the average physical layer throughput, R_{avg}. The SDR supports two waveforms for good and bad channel conditions with rates R_{high} and R_{low}. When the instantaneous throughput drops to R_{low} due to a fade, data will start getting backed up. The buffer must be large enough to hold $(R_{avg} - R_{low})$ for the duration of the fade, as shown in Fig. 4.1.

---

[8] Since FPGAs are used in most SDRs today, a hardware virus may be able to physically damage the radio by overheating the device [268]. The feasibility of such a virus in modern FPGAs is unknown.

[9] This discussion does not apply to data sources that can vary the rate based on feedback from the physical layer (see, for example, Sect. 3.7.2.2).

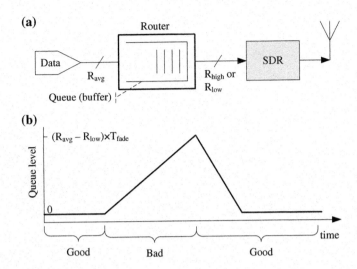

Fig. 4.1 Router buffer size requirement to support ACM. **a** Conceptual view, **b** queue level versus time

Since the fading channel is random, fade duration is theoretically unlimited. The buffer must be sized such that data is not lost more than some specified percentage of the time (e.g., no more than 0.01 %). The average throughput and buffer size can be derived analytically by assuming a Markov model for transitions between different ACM waveforms [37]. Transition probabilities are derived by analyzing long-term channel statistics.

Chapter 5
Signal Processing Devices

Transistors are cheap and getting cheaper. This key observation has made SDR a reality. A true SDR has to implement at least some of the physical layer functionality in software. For the purpose of this chapter the definition is expanded to include software-controlled radios. Flexible and programmable digital signal processing can be implemented on a wide range of devices. This chapter provides an overview of the different types of devices suitable for implementing the DSP portion of an SDR.

5.1 General Purpose Processors

A general purpose processors (GPP) is a typical microprocessor, like the ones powering personal computers.[1] As the name implies, these devices are optimized to handle the widest possible range of applications. A GPP must perform well for browsing the web, word processing, decoding video, scientific computations, etc. The very disparate tasks preclude domain-specific optimizations. GPPs must excel at fixed and floating-point operations, logical operations, and branching. This makes them suitable for implementing much of the SDR functionality, starting with the physical layer DSP and ending with the protocol and network stacks. GPPs provide the easiest development environment and highest developer productivity. The largest pool of qualified developers is familiar with GPPs. The development tools and expertise apply to GPPs from different vendors and product lines. A wide range of operating systems is available, from full featured graphical interfaces such as MS Windows to lean real-time products such as VxWorks. These advantages make GPPs the easiest platform for SDR development as will be discussed in Sect. 6.1.

---

[1] Major vendors include Intel, ARM, AMD, MIPS.

E. Grayver, *Implementing Software Defined Radio*,
DOI: 10.1007/978-1-4419-9332-8_5, © Springer Science+Business Media New York 2013

5.2 Digital Signal Processors

A digital signal processors (DSP) is a microprocessor that is optimized for number crunching.[2,3] Manufacturers of DSP devices can optimize them for a much narrower set of target applications than GPPs. Power consumption can be reduced by eliminating transistors that a GPP has to devote to sophisticated cache and peripheral subsystems. The main advantage of DSPs over GPPs is in power consumption per operation. DSPs come in both fixed- and floating-point varieties, with fixed-point versions offering even lower power consumption. DSPs are not well suited for control—(rather than datapath)—intensive code such as the protocol and network stack. A typical SDR would pair a DSP with a GPP to implement the network stack.

The development environment for a DSP is somewhat more complex than for a GPP. Operating system support is also significantly more limited, with many DSP projects not using any OS at all and interacting with the hardware directly. If an OS is required, a real-time one is almost always used. DSP developers are significantly more difficult to find than GPP developers. Optimal use of a DSP requires the developer to be very familiar with the internal architecture, and the expertise in one family does not translate to another family.

DSPs are used extensively in software-defined cellular base stations and in radios that require low power and have modest data rate requirements. In the author's opinion, DSPs are a niche product and do not offer a compelling advantage over either GPPs or FPGAs.

5.3 Field Programmable Gate Arrays

A field programmable gate array (FPGA) is a microchip that is designed to be configured by the user after manufacture. An unconfigured FPGA has absolutely no functionality.

FPGAs contain programmable logic components called "logic blocks" and reconfigurable interconnect to "wire" the blocks together—somewhat like a one-chip programmable breadboard (Fig. 5.1). Logic blocks can be configured to perform complex combinational functions or simple logic such as AND, NOT, XOR, etc. Each logic block is usually paired with one or more registers to store the result. Most FPGAs also include large "macro blocks" for frequently used functions such as memory blocks, multipliers, interfaces to external devices, and even complete microprocessors. The macro blocks could be usually built up from the logic blocks, but the resultant design would be slower and use up a lot of the

---

[2] Major vendors include Texas Instruments, AMD, Free scale.

[3] There is unavoidable acronym confusion when discussion digital signal *processors* (DSP) used for digital signal *processing* (also DSP). The second meaning is used more frequently.

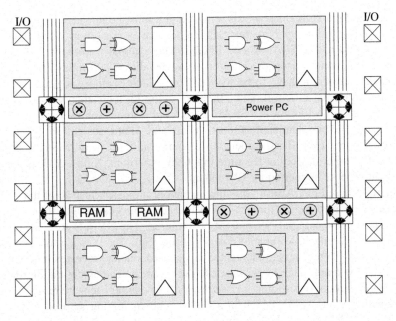

Fig. 5.1 Conceptual block diagram of an FPGA, note the programmable interconnect using cross-bar connections

available logic blocks. FPGA vendors attempt to differentiate themselves by providing the *right* mix of logic and macro blocks.[4]

The interconnect can be used to implement the desired functionality. For example, a few gates can be combined to implement an adder, while a few more implement a multiplier. However, the overhead of all the wiring and switches makes FPGAs consume significantly more power ($10\times$) than an equivalent single-function design. Designs implemented on an FPGA also execute a lot slower ($5\times$) than an equivalent single-function design [38].

Three types of FPGAs can be defined based on the way they are configured:

1. SRAM-based FPGAs use volatile memory registers to configure the routing between logic elements and the detailed logic element implementation. The

---

[4] For example, a high-end FPGA from Xilinx in 2011 (XCE7V200T) provides the following set of blocks. Vendors are still experimenting with providing complete microcontrollers as dedicated blocks inside an FPGA.

Logic gates	Registers	RAM blocks	Multipliers (w/adders)	Gigabit transceivers	Clock generators
2 M	2.5 M	6 MB	2,100	36	24

configuration is lost if power is removed and must be reloaded every time the device is powered up. Note that this implies that external, non-volatile memory is required to use SRAM-based FPGAs. The external memory may reside on the same circuit board, or could be on a hard disk 1,000 miles away, but the configuration data have to be loaded. The volatile memory registers are very small and leave a lot of room on the FPGA chip for logic elements and routing. Thus, all large FPGAs are SRAM-based and account for most of the FPGA market. The two dominant vendors, Xilinx and Altera, account for 90 % of all FPGA sales [39]. Their products are quite similar,[5] with Xilinx providing slightly larger devices at the time this book is written. Xilinx devices are also slightly better supported by IP vendors.

2. FLASH-based FPGAs use non-volatile memory registers instead of the volatile ones. One advantage is that the FPGA is ready to process data as soon as power is applied; no external NV memory is required. The NV registers consume no power resulting in lower idle power consumption (by as much as 2×). However, FLASH registers implemented in standard silicon processes used to make FPGA logic are much larger than SRAM registers. The configuration registers leave less room on the chip for user logic.[6] FLASH registers are more robust to radiation effects [40], which is a major concern for the aerospace industry. Some SRAM-based FPGAs include FLASH memory either in the same package (e.g. Xilinx Spartan-AN) or even on the same die (e.g. Lattice XP2). However, such devices do not have the radiation tolerance advantages of a true FLASH FPGA. The only FLASH FPGA vendor is Microsemi (previously Actel) [41].

3. Antifuse-based FPGAs rely on unique chemical switches between interconnects. These switches are normally open, but can be closed by passing current through them. Once closed, the switch cannot be opened again. Thus, antifuse FPGAs can be programmed only once. Antifuse-based FPGAs are therefore not suitable for SDR implementation because the functionality cannot be changed. Once programmed, they should be considered ASICs. Antifuse switches are even larger than FLASH configuration registers, leaving little room for user logic.[7] Antifuse FPGAs are used almost exclusively by the aerospace industry because of high radiation tolerance. The only antifuse FPGA vendor is Actel [42].

To the best of the author's knowledge, only SRAM-based FPGAs have been used to implement SDRs for both terrestrial and space[8] use. The high signal

---

[5] Each vendor has a devoted following who would disagree with this statement.

[6] The largest flash FPGA in 2011 is the Actel RT3PE3000L with approximately 75 K registers. Compare that to the 2.5 M registers in a top-of-the-line SRAM FPGA.

[7] The largest antifuse FPGA in 2011 is the Actel RTAX4000S with approximately 10 K registers.

[8] A radiation-tolerant antifuse FPGA is used to control and configure the SRAM FPGAs in space-based SDRs. SRAM FPGAs are used for all the signal processing. A new radiation-tolerant SRAM FPGA was recently developed by Xilinx and the US Air Force [263].

processing requirements of a modern SDR simply exceeds the capabilities of FLASH devices.

The ability to implement *exactly* the functionality required for the radio[9] without any unused logic makes FPGAs dramatically more efficient than GPPs.

FPGAs excel at heavy-duty number crunching in a datapath flow, but are not as well suited for control-intensive processing. A microcontroller is often implemented using some of the logic gates or an external microcontroller is paired with the FPGA. The microcontroller is then used for the control-intensive processing. The development flow and environment for an FPGA are very different from those used for DSP or GPP (see Sect. 11.3).

5.4 Specialized Processing Units

A number of devices suitable for DSP do not fall neatly into any of the categories above. These are referred to as SPUs and will be called out by name when necessary. Specialized processing units (SPUs) typically combine multiple cores with different characteristics to allow efficient mapping of SDR algorithms. An interesting overview of some modern SPUs developed specifically for SDR is provided in [43].

The GPP used to be the most complex and powerful chip in a PC. Today the graphics processor (GPU) responsible for displaying images on the screen is often more complex than the GPP. Image processing can be expressed as a set of linear algebra operations. Much of the signal processing required for the physical layer of an SDR can be expressed using similar operations. A GPU is essentially an array of floating-point multipliers with efficient access to memory and can be considered a very specialized DSP. In terms of raw performance (GFLOPs), and single GPU outperforms a GPP by a factor of at least 5 and a DSP by a factor of 3. The increased specialization makes them more efficient but more difficult to program. GPU vendors have realized the potential new market for their chips and each new generation is becoming easier to program for non-imaging applications. General purpose GPUs (GPGPUs) are slightly tweaked GPUs marketed to high-performance computing users.

The most mature GPGPU programming language, CUDA, is developed by NVIDIA [44]. An open source language, OpenCL, is gaining ground [45, 46]. Writing software to take advantage of these devices remains very challenging because of non-intuitive interaction between processing elements and memory. Optimized function libraries can be used to take advantage of GPGPUs from

---

[9] The designer can choose to use either fixed- or floating-point logic, at any precision. For example, if the input data is only 1 byte (8 bits) a GPP still has to use the 64-bit adder to process the data. In contrast, an FPGA can be configured to implement an 8-bit adder, making it more efficient.

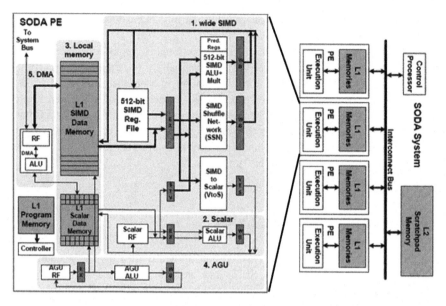

Fig. 5.2 SODA processor architecture [50]

popular development environments such as MATLAB and C ++ without having to program in CUDA of OpenCL (e.g. [47]).

An obvious question is: If GPUs are so fast, why are there no similar devices optimized for SDR instead of video processing? The answer is cost. GPUs are relatively inexpensive because of the huge volume—a lot more people are buying high-end video cards for gaming than are buying SDRs. Since SDR is still a relatively small market, it has to either use general purpose devices (GPP, DSP, FPGA) or repurpose devices developed for larger markets.

The IBM Cell processor [48] is another device that has been successfully used for SDR development. Cell was developed for image processing required for video games. However, the architecture was developed from the ground up to support applications other than video games. Unlike GPUs, a Cell processor includes a full featured GPP in addition to an array of specialized DSPs (synergistic processing element, SPE). The Cell processor also offers lower power consumption per GFLOP than a GPU. However, since GPUs are sold in much larger volumes, the Cell processor is much more expensive.[10] A robust general purpose (rather than graphics specific) development environment was available for the Cell from the beginning. This made it an attractive target for HPC and even motivated a version of the popular open source SDR framework (GNURadio, see Sect. 8.1).

---

[10] The only inexpensive Cell processor system is a Sony PS3 because it is subsidized by the manufacturer.

The SODA architecture [49] (and the derived Ardbeg chip) was designed specifically for SDR [50]. This chip can be considered a scaled down version of the Cell processor with a GPP augmented by four SIMD cores. The Ardbeg chip was never released commercially. The architecture, shown in Fig. 5.2, is representative of a class of SIMD-augmented processors. A nice summary of Ardbeg features and comparison to similar processors is provided in [50].

A host of other devices developed for flexible signal processing can be used to implement an SDR. There is an interesting intersection between chips targeted at the SDR and the HPC markets. Both markets are rather small and somewhat crowded with established players. This makes new entrants likely to fail, and makes adoption of these new devices risky. The specialized nature of new devices implies a significant learning curve to effectively use the resources and the development flow. Investment in time has to be carefully weighed against the advantages of the device given the risk of it disappearing in a year or two [51]. Some of the notable failures in the last decade include: morphX, mathchip, sandbridge [52, 53]. Some of the remaining entrants are:

- Coherent Logic's HyperX, which claims to offer both low-power consumption and high-computational throughput [54]. The architecture is somewhat similar to that of a GPU.
- Octasic offers a multi-processor solution with each processor derived from a DSP (rather than a GPP) [55].
- picoChip[11] offers arrays of up to 273 primitive DSPs [56].
- Tilera offers yet another massively multi-processor device [57]. The Tile processor is representative of the other devices mentioned above and will be discussed in more detail below.

5.4.1 Tilera Tile Processor

The trend in both high-performance and low-power computation is to integrate multiple processor cores on a single die. Simply increasing clock rates has become infeasible, mostly due to heat dissipation issues. Modern high-performance processors offer 6 or more fully functional cores on a single die. The embedded, low-power processors are pushing the number much higher. Tilera markets a 64-core device, which is about half way between the 4+ core Sandbridge and the 270+ core picoChip. As mentioned previously, many of the companies offering novel processor architectures are very short lived. The Tile64 processor from Tilera received a strong vote of confidence from the military space community

---

[11] As this book went to print, picoChip was acquired by Mindspeed Technologies and the future of its products is uncertain.

Fig. 5.3 MAESTRO chip
architecture **a** top level,
b detail of one tile

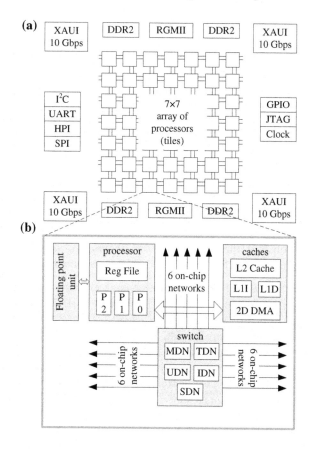

when Boeing was commissioned to develop a space-qualified version [58]. The
radiation hardened by design (RHBD) version of Tile64 is called MAESTRO.
Space-qualified chips have much longer life cycles than their terrestrial equivalents. A payload designed around the MAESTRO chip may take 5 to 10 years to
complete and launch. Radiation hardening is very expensive and the few chips that
are available get used for multiple programs. Entire sessions at multiple space-
related conferences are now devoted to the MAESTRO chip [59, 60]. Developers
can feel relatively confident that software tools will remain available even if Tilera
does not succeed in the marketplace.

The MAESTRO chip is built around a 7 × 7 array of processors,[12] known as
tiles, as shown in Fig. 5.3. The 49 tiles share access to four memory interfaces and
four high-speed serial interfaces. Each MAESTRO tile is a three-way VLIW
processor with 8 kB L1 data cache, 64 kB data/instruction cache, a DMA engine,

---

[12] Note that it is smaller than the commercial counterpart. Commercial Tile processors do not
have a floating-point unit.

and a floating- point unit. Each tile uses about 200 mW when running at 300 MHz. The peak performance for the entire chip is 44 GOPs and 22 GFLOPs.

Tiles are interconnected in a two-dimensional mesh with five high-speed, low latency networks between every four tiles (and one system network). Each tile includes a switch to route packets. Note that latency increases linearly with the distance between the source and destination tile for a packet. Software algorithms should therefore be designed to minimize data transfers between non-adjacent tiles.

One of the main strengths offered by Tilera is ability to run standard Linux and VxWorks operating systems. This means that porting an SDR framework such as GNURadio (see Sect. 8.1) to Tile64 is relatively straightforward. A direct port would not make efficient use of the cores, but allows gradual optimization. Mapping each SDR functional block to one or more adjacent tiles is a simple optimization step. However, best performance can only be achieved by leveraging the hand-optimized libraries provided by Tilera.

5.5 Application-Specific Integrated Circuit

Application-specific integrated circuit (ASICs) are not typically mentioned in the context of SDR, but remain a viable alternative for very flexible radios.[13] An ASIC can be used to implement a software-controlled radio. ASICs power most of today's cell phones and other handheld devices and allow them to support multiple waveforms. The ASICs in cell phones implement a system-on-a-chip (SoC) with one or more GPPs and DSPs in it. The most numerically intensive functional blocks are implemented in dedicated circuits or accelerators, while the rest of the processing is done in software. ASIC development flow is quite similar to FPGA development flow, with additional steps required for the final manufacture.

5.6 Hybrid Solutions

As transistor density keeps increasing, manufacturers can put multiple devices on a single chip. A number of hybrid devices are available that combine two or more of the processors described above. One of the first commercially available examples was the OMAP architecture from Texas Instruments. An OMAP chip combines a GPP with a DSP [61].[14] The ZYNQ family of devices from Xilinx combines

---

[13] Technically, all of the processor types described above are themselves ASICs (i.e., an FPGA is a large ASIC that happens to be programmable).

[14] The GPP/DSP architecture has been abandoned by TI in favor of multiple GPPs and dedicated hardware accelerators.

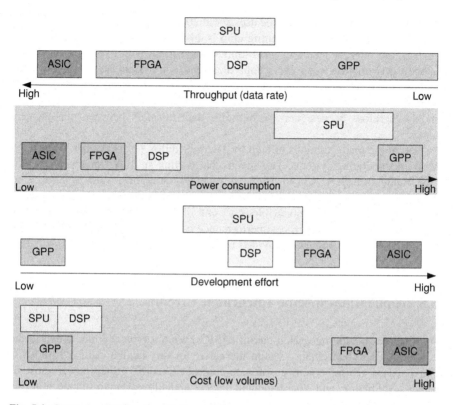

Fig. 5.4 Comparing DSP implementation options

a GPP with an FPGA [62]. Hybrid solutions do not introduce any new capabilities but tend to reduce product cost and size.

5.7 Choosing a DSP Solution

A qualitative comparison of the different DSP options is shown in Fig. 5.4.
    The following conclusions can be reached by examining that figure:

- ASICs are pretty much the only option that can achieve high throughput at reasonable power consumption. It is hardly surprising that ASICs are used in almost all high-volume handheld consumer devices. Modern ASICs often include a GPP and thus offer a limited level of programmability.
- FPGAs should be considered for high-throughput applications that can tolerate somewhat higher power consumption. The latest devices are dramatically more power efficient than just two years ago. As the power consumption of the DSP becomes negligibly relative to the RFFE, FPGAs become an obvious choice.

- GPPs can now process low- to-medium throughput signals in real time. They offer unmatched flexibility and ease of development. GPPs should be the first choice for SDRs when power consumption is not a major concern.
- DSPs are definitely niche devices. They do not offer sufficiently higher throughput or lower power than alternative options. They are used when GPPs are either not fast enough or consume too much power *and* the development team does not have FPGA expertise.
- SPUs must be considered on a case-by-case basis. The two key concerns when considering SPU are device longevity and ease of development.

Three major signal processing architectures can be defined based on Fig. 5.4:

- All-software using a GPP
- GPP with hardware acceleration using an FPGA or SPU
- FPGA

Note that the DSPs are not used in any of these architectures. This omission is partly due to the author's bias against DSPs—they are only marginally easier to work with than FPGAs and almost always offer lower performance (for an interesting comparison, see [63]). The optimal architecture depends on many factors, including:

- Expected signal processing requirements (in operations per second[15])

  - The bandwidth of the signal(s) to be processed is a rough indicator of the required signal processing. A wideband signal is either high data rate or highly spread. In the first case, the FEC will require a lot of processing, while in the second case the DSSS acquisition will.

- Number of signals. Assuming all the signal(s) of interest are digitized together, rather than separated in the RF front end, the DSP has to handle the entire bandwidth before separating the signals for individual processing. Wide input bandwidth makes it difficult or impractical to use a GPP for initial channelization. On the other hand, it is more difficult to develop FPGA firmware to simultaneously support multiple narrowband signals than a single wideband signal.
- Operating environment. All of engineering is about achieving the desired goal given a set of constraints. The constraints for an SDR radio used in a research/lab environment are very different from those for a handheld consumer device. Three classes can be defined to express different target environments:

---

[15] An operation can refer to a floating point, fixed point, or bit operation. A GPU/SPU/DSP uses whatever arithmetic and logic unit (ALU) is available for each operation, while an FPGA can implement each operation in a dedicated circuit. Thus, comparing the DSP capabilities based on GFLOPs or GMACs is only often misleading.

	Size	Power	Cost
Rack	Size is essentially unconstrained	Power is essentially unconstrained, only limited by heat dissipation to <10 kW	Cost is not a driver
Desktop	Size is constrained to a volume of about 1 m^3—a large PC tower or test instrument.	Power is constrained to a single outlet <1 kW	
Handheld	Size is a key design driver and should be as small as possible, but no larger than 100 cm^3.	Power is just as important as size, limited to <1 W on average	Cost is sensitive for high-volume devices.

- Flexibility and responsiveness. A consumer device may never need to change the supported waveforms after it has been shipped, or the updates may come once a year. Conversely, a research SDR may require different waveforms to be implemented on a weekly basis. The level of effort and manpower required to implement changes can be a deciding factor.
- Available skill set. If multiple architectures satisfy the 'hard' constraints, the final decision frequently hinges on the development team's skill set. A longer term view should also take into account the ease of bringing new/additional developers into the group. A single expert[16] in a unique device or tool chain can steer a group down a path that results in a steep learning curve and lost productivity.
- Maintainability. A development effort with a well-defined end date and final goal can take advantage of the current state-of-the-art devices. However, a long-term strategic development effort must consider how the selected device will evolve over time. An SPU from a small start-up can be a perfect fit for the task at hand, but may not offer a clear evolutionary path. Likewise, development for the current FPGA must consider how the unique capabilities offered by that device may limit the value of the developed codebase once the next generation of the FPGA is available.[17]

  – The GPP is indisputably the best platform for maintainability. Well-written[18] C++ or Java code from 20 years ago can still be compiled today.

---

[16] This problem is especially acute in large corporations with a lot of institutional knowledge. See the discussion of SPW and Simulink in Chap. 11.

[17] Maintainability and ease of upgrades are strongly affected by the development toolset. The higher the abstraction level, the easier the upgrade path … unless the toolset used to convert from abstract description to the implementation disappears. See the discussion of design flows in Chap. 11.

[18] Unfortunately, writing portable code often means eschewing the use of optimized performance libraries, since the libraries may not be available or may change significantly in the future.

Chapter 6
Signal Processing Architectures

One or more devices described in Chap. 5 can be used to implement an SDR. This chapter considers SDR architectures based on each DSP option. The receiver in an SDR almost always requires significantly more computational resources than the transmitter. Thus, only receiver architectures will be discussed in this chapter.

6.1 GPP-Based SDR

A GPP-based SDR offers maximum flexibility and the easiest development. GPPs are best suited for SDRs used in a laboratory environment for research and development, because size and power consumption are not significant concerns and the ability to quickly try out a range of algorithms and waveforms is critical.[1] The maximum data rate that can be supported on a high-end PC in 2010 was about 1 Mbps for a reasonably complex waveform (see Sect. 8.1.4 and [64]). This data rate can be achieved without heroic measures by parallelizing the signal processing. The PC receives digitized baseband or low-IF samples over one of the standard interfaces (Ethernet, USB, FireWire, PCI). The samples are buffered and passed to the signal processing threads.

Latency is a major concern for GPP-based SDRs. GPPs achieve high performance only when working on blocks of data, rather than on one sample at a time.[2] The operating system also introduces latency and jitter when handling interrupts and context switching for different threads. The jitter can exceed 10 ms on a Windows system but a real-time operating system such as VxWorks can achieve

---

[1] Real-time operation is assumed. A nonrealtime (offline) software-defined radio is best characterized as a simulation platform.

[2] A single-instruction-multiple-data (SIMD) floating point unit is available on modern GPUs from Intel and AMD. It can perform the same operation on up to eight samples at once. However, overhead to set up the SIMD transaction is only justified for blocks larger than 128 samples. In practice, blocks of thousands of samples are used to reduce overhead. Note that very large blocks that exceed the cache size cause performance to drop.

E. Grayver, *Implementing Software Defined Radio*,
DOI: 10.1007/978-1-4419-9332-8_6, © Springer Science+Business Media New York 2013

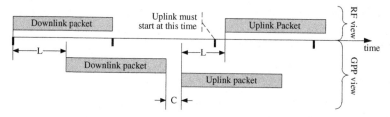

Fig. 6.1 Missing the uplink slot due to excessive latency

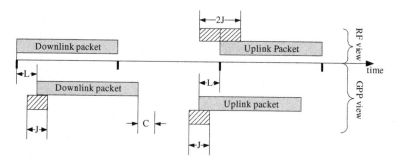

Fig. 6.2 Missing uplink slot due to excessive jitter

jitter below 1 ms. Jitter and latency due to the digitizer interface must also be considered—under 10 μs for 10 GbE [65], under 100 μs for 1 GbE [66], up to 200 μs USB [67]. Jitter and latency become a major concern if the waveform requires the transmissions to be scheduled based on received data. Many commercial standards (e.g. WiMAX, UMTS) place this requirement, and the turn-around time is often tight (<10 ms). If the latency from the digitizer to the software running on the GPP exceeds half[3] of the turn-around time, the waveform cannot be supported. Likewise, if the jitter exceeds half of the uncertainty window, the waveform cannot be supported. An example showing the effect of latency is presented in Fig. 6.1. A downlink packet is transmitted at a slot boundary in a TDMA system (see Sect. A.5.1). It is received by the GPP after L seconds due to latency. Some time, C, is required to process the received packet and generate a response. The response packet reaches the antenna another L seconds later and misses the start of the time slot.

The effect of jitter is shown in Fig. 6.2. In this scenario, latency is small enough to give GPP plenty of time to prepare the uplink packet. Uplink is started L seconds before the uplink time slot. On average, the uplink packet will be

---

[3] Latency is incurred when receiving and again when transmitting the response.

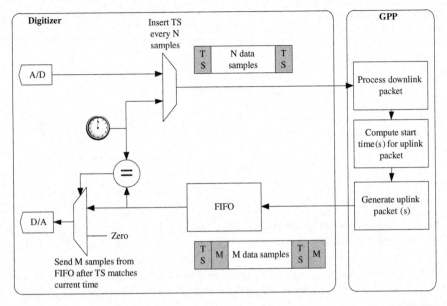

Fig. 6.3 Possible implementation of a time-aware mixed-signal front end for a GPP-based SDR (*TS* indicates a timestamp)

transmitted at the correct time. However, the combined receive and transmit jitter, $2J$, may cause the start of the uplink packet to fall outside the allowed uncertainty range (i.e. it could start J seconds earlier or later than desired).

Some waveforms require accurate synchronization to absolute (e.g. GPS) time rather than to the received signal. For example, a spreading code in a DSSS waveform can be a function of GPS time (see Sect. A.5.5). The code cannot be acquired without knowing the time, and the acquisition complexity grows linearly with time uncertainty. Relatively accurate time[4] can be acquired on a PC using either a dedicated GPS-based timing card, or network-based synchronization tools such as NTPv4 [68] or IEEE 1588 [69]. However, the accuracy is still limited by the jitter of the digitizer to PC interface. The digitizer can be configured to add timestamps to the data samples, as shown in Fig. 6.3. The timing accuracy is then only limited by the digitizer hardware and can be made very accurate ($\ll 1$ μs) if the digitizer has access to GPS time and a 1 pulse per second signal.[5] Accurate scheduling of transmissions can be greatly simplified if the hardware is responsible for starting the transmission. The hardware has to have a buffer to hold samples

---

[4] A well implemented IEEE 1588-2008 on a Linux machine with a time server.

[5] Access to the absolute (e.g. GPS) time is usually not required. A 1 PPS signal can be used to simultaneously reset a free-running counter and any internal buffers. The GPP can then correlate the local time acquired by any means with received data samples. Local time has to be accurate to 1 s. Software uses the nominal sample rate to compute the time of arrival for the Nth sample.

Fig. 6.4 Nonrealtime short
burst transmitter using a
low-power GPP

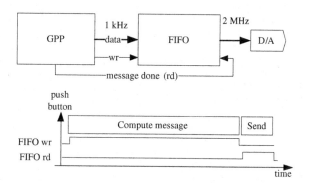

before transmission. The GPP can use timestamps[6] from the received data to compute the appropriate transmit time for each packet. In fact the GPP can generate multiple packets and schedule them for future transmission, greatly simplifying the software. For example, each packet can be prefixed with a header containing the desired transmit time and the length of the packet, as shown in Fig. 6.3.

The time-aware front end eliminates any timing jitter caused by the interface to the GPP. However, it does not help with latency.

6.1.1 Nonrealtime Radios

Most radios operate in real-time, or close to real-time. That is, the radio is always processing received samples and the processing rate must be close to the sample rate (otherwise the radio falls behind and can lose data). In the author's experience, whenever SDR is mentioned, the real-time operating requirement is all but assumed. There is a nonnegligible class of radios that can operate much slower than the sample rate required to generate or capture the waveform.

Consider the humble garage door opener. Some of the modern systems use a relatively wideband, spread-spectrum waveform. A very short burst is transmitted from a key fob after the button is pressed. The message changes every time to prevent record-and-playback attack. The user does not care if the burst is sent a second or two after the button has been pressed.

A key fob for the example above should be small and low-power. However, a low-power GPP cannot generate a contiguous waveform at the required chip rate (e.g. 1 Mcps). One alternative is to use a more powerful GPP or even go to an FPGA implementation. Consider a hypothetical system that transmits a 10 bit

---

[6] Note that for the TDMA system described above, the absolute time is not necessary because transmit time is computed based on the time a downlink packet was received.

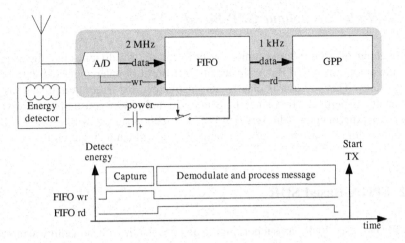

Fig. 6.5 Non-realtime short burst receiver using a low-power GPP

message at a rate of 1 Kbps, spread with a 1 Mcps code. The duration of the message is 10 ms = 10,000 chips = 20,000 samples (assuming 2 samples/chip). A low-power GPP can compute the required samples and put them into a 20 kB (assuming 8 b/sample) FIFO[7] at some arbitrarily low rate (e.g. 1 kHz). Once the message is complete, the FIFO can be read out at the required 2 MHz rate, as shown in Fig. 6.4.

Design of a low-power nonrealtime receiver is less obvious. The transmitter was started by the user pressing a button. A typical receiver has to be always active since the message can arrive at any time.[8] One option for nonspread waveforms is to rely on a passive power detector to start capturing the signal. However, the SNR for a spread signal is too low to trigger a power detector. A nonrealtime transceiver is also suitable for battery assisted passive (BAP) RFID tags [70]. Incident RF energy from the interrogator, captured by a passive energy detector, is used to wake up the transceiver (Fig. 6.5). Once awake, the receiver captures a block of samples into a FIFO. A low-power GPP can then process the captured samples at low speed. The transmitter operates similar to the example above, with the receiver providing the trigger instead of the push button.

---

[7] Some microcontrollers have enough on-chip memory to implement this FIFO. The microcontroller would also have to provide a synchronous external interface that can be directly interfaced to a DAC chip.

[8] Some wireless systems define regular intervals at which the receiver must be woken up to check if a signal is present (e.g. every minute, on the minute). This approach requires the receiver to keep track of absolute time.

6.1.2 High-Throughput GPP-Based SDR

Throughput of a GPP-based radio can be increased by dividing the signal processing among many GPPs. A cluster of GPPs must be interconnected with high speed links. This solution is clearly only applicable for large ground stations or laboratory experimentation since both the size and power consumption are very high. An experimental radio reported in [71] supports signals with up to 40 MHz of bandwidth, at least an order of magnitude higher than a single GPP.

6.2 FPGA-Based SDR

An FPGA-based SDR can support high data rates and high bandwidth waveforms in a relatively compact and low-power package. At the time this book was written, high-end FPGAs provide enough signal processing resources to support most practical waveforms in a single device. Multiple devices may be required for an unspread waveform with a bandwidth above 300 MHz or a spread waveform with a large acquisition space.[9] Low-end FPGAs are suitable for bandwidths from 10 MHz down to 100 kHz, and provide much lower power consumption than their high-end counterparts. Although FPGAs can be used to process narrowband waveforms below 100 kHz, other devices such as DSPs and GPPs may be a better alternative.

As discussed previously, FPGAs are fundamentally different than GPPs. A GPP executes instructions from memory and can easily switch from one function to another. Additional functions do not slow down the processor or require more resources.[10] An FPGA has all the functionality mapped to dedicated circuitry, and each new function requires more FPGA resources. Despite this difference, FPGAs can be used to implement very flexible radios and support a wide variety of waveforms. There are four ways (discussed in next subsections) to support different waveforms on an FPGA:

1. Create separate configuration files for each waveform and load them as needed.
2. Develop the firmware to support all the waveforms in a single configuration.
3. Develop a small set of configurations to support different waveform classes and support multiple waveform variations within a class.
4. Take advantage of partial reconfiguration offered by some FPGA devices to load only the difference between waveforms.

---

[9] The DSSS acquisition problem can be made arbitrarily difficult by extending time and/or frequency uncertainty..

[10] Other than memory to store the instructions, which can often be neglected.

6.2.1 Separate Configurations

Unlike a GPP, it takes time for an FPGA to load a new configuration.[11] A large FPGA can take over 100 ms to load.[12] The configuration files are also quite large, up to 40 MB per configuration. Reloading the FPGA causes it to lose all state information, and an external memory is required if data must be saved between different configurations. This approach gives the designer the greatest flexibility for designing firmware to target a specific waveform. All the FPGA resources are available for just that waveform, making the best use of the FPGA. However, the number of configurations can quickly become impractical.

6.2.2 Multi-Waveform Configuration

The firmware is designed to support all the expected waveforms. Unique features of every waveform are selected by setting appropriate registers at run time. This approach places a much greater burden on the developer and makes the code significantly more complex. For example: a continuously variable resampler is required to support different data rates; tracking and acquisition algorithms must support different waveforms; etc. Implementation of FEC codecs for different algorithms is typically so different that a dedicated implementation is required for each.[13] FEC decoders are typically the most complex and resource intensive blocks in a radio receiver, and having dedicated implementations for each can exceed the available resources on an FPGA. Runtime reconfigurable logic is always slower than single-function logic. Thus, firmware designed to support multiple waveforms achieves lower throughput.

The major advantage of this approach is the ability to switch between waveforms almost instantaneously. Rapid reconfiguration is required for some waveforms (see the discussion of ACM in Sect. 3.1.1). Another potential advantage is that the firmware for a multi-waveform configuration can be retargeted from an FPGA to an ASIC. An ASIC implementation is justified if the SDR is used in a high volume, low power, and cost-sensitive product.

Many waveforms are very similar to each other, and only minimal changes are required to the firmware for one to support the others. For example, if the

---

[11] An interesting FPGA architecture from Tabula [286] reconfigures the FPGA almost instantaneously. However, the reconfiguration is not under user control and is meant for resource reuse rather than real-time functionality change.

[12] Virtex XC7V1500T requires about 40 MB of configuration data and can be loaded over a 32-bit bus at 60 MHz.

[13] Reconfigurable codecs can be developed to cover some algorithms [281].

data rate changes between two waveforms, a simple decimator/interpolator is required.[14] An interpolator uses few FPGA resources and is easy to implement. On the other hand, a GSM waveform shares almost nothing with a WiMAX waveform. Grouping waveforms into classes based on similarity of signal processing motivates a hybrid approach with each configuration supporting an entire class of waveforms.

6.2.3 Partial Reconfiguration

Some FPGAs[15] allow only part of the device to be reconfigured, while the rest of the device continues operating. This technique is known as partial reconfiguration (PR) [72] and has been available to designers for almost a decade. Unfortunately, the implementation of this approach is rather difficult and suffers from limited development tool support. It is easy to come up with plausible applications of partial reconfiguration, but the advantages are less compelling than they first appear.

Partial reconfiguration supported by Xilinx is coarse grained—i.e. relatively large sections of the device have to be reconfigured at once. At one extreme, PR reduces to the separate configurations approach, with only a small section of the FPGA retained between configurations. A major main advantage of 'almost complete reconfiguration' is that an external processor is not needed to manage the configurations. Instead, an embedded processor is used. The embedded processor is also responsible for retaining state information from the reconfigured region.

Partial reconfiguration fits well with hybrid FPGA/GPP architectures (see Sect. 6.5) as well as with hardware accelerated architectures (see Sect. 6.6). The excellent work in [73] provides a mathematical framework to schedule the loading of reconfigurable segments onto an FPGA to accomplish the desired computation.

SDR is an application always mentioned by FPGA vendors when touting PR features of their latest devices. The next few pages explore a few specific examples [74].

6.2.3.1 Partial Reconfiguration for ACM

A radio using adaptive coding and modulation (ACM) (see Sect. 3.1) chooses different waveforms to efficiently use available spectrum under varying channel conditions. Partial reconfiguration appears to be an elegant solution for supporting

---

[14] For example, the WiMAX standard defines essentially the same waveform for channel bandwidth of 3.5, 5, 7 and 10 MHz.

[15] Both major vendors, Altera [283] and Xilinx [282], support this technique..

Fig. 6.6 FPGA with one
fixed and two reconfiguration
regions (RR)

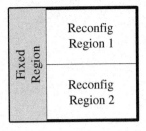

Fig. 6.7 ACM
implementation using partial
reconfiguration

| ACM | 16-QAM demodulator | | ACM | BPSK demodulator |
| TX | Viterbi (r=¾) decoder | | TX | LDPC (r=¼) decoder |

(a)   (b)

fast ACM in small FPGAs. Consider an FPGA partitioned into one fixed region
and two reconfigurable regions as shown in Fig 6.6.[16]

Figure 6.7 shows one possible scenario where the receiver is initially configured for a high-throughput waveform. The transmitter, implemented in the fixed
region of the FPGA, sends channel SNR updates to the other radio. When the SNR
decreases, the ACM algorithm executing on the other radio initiates a waveform
change. The waveform change command is processed by the ACM block (in the
fixed region of the FPGA). ACM block then initiates reconfiguration. The first
region is reconfigured to implement a demodulator for a robust waveform (BPSK).
The second region is reconfigured to implement a more powerful and lower rate
FEC decoder.

Consider a fast-ACM system in which every frame carries a preamble to
indicate the code and modulation (CMI) used in the payload (see Sect. 3.1.1). It is
critical that the time required to reconfigure the FPGA and then process a frame of
data is less than duration of two frames. Incoming data are buffered while the CMI
is parsed in the fixed region. Logic in the fixed region or an external GPP then
loads configuration into RR1 to process CM1. Once CM1 is loaded, buffered data
are processed. In the meantime, CMI for the second frame has been parsed by the
fixed region and RR2 is loaded. The pipeline continues with sufficient throughput
to support contiguous data (Fig. 6.8).

---

[16] The fixed part of the FPGA is shaded in all the figures in this section, and the reconfigurable
part is not shaded.

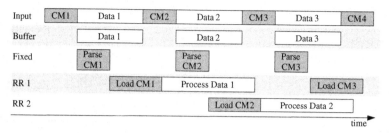

Fig. 6.8 Timing for fast ACM using partial configuration

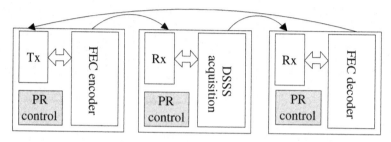

Fig. 6.9 Simplex DSSS transceiver with FEC

Consider a DVB-S2 waveform that requires one-fourth of a relatively large FPGA. Reconfiguration time is on the order of 5 ms.[17] If the data rate is 5 Mbps (one HDTV channel), each 16 kb frame takes about 3 ms. Two frames take 6 ms, which is just over the reconfiguration time. However, if the data rate is 50 Mbps, each frame takes 0.3 ms. At least $(5 \text{ ms}/ (2 \times 0.3 \text{ ms})) \approx 9$ reconfigurable regions would be required to maintain the throughput. Unless the number of CMs is very large *and* each CM is very different from all others, one configuration that supports all the CMs will require fewer FPGA resources than nine reconfigurable regions.

6.2.3.2 Simplex Spread-Spectrum Transceiver with FEC

One potential SDR use would be in a simplex transceiver, where only transmit or receive capabilities are used at any time. Assuming that the waveform requires FEC as well as direct sequence spread spectrum (DSSS), the design could be organized as shown in Fig. 6.9. Two PR regions are declared: one for either the Tx modulator or the Rx demodulator, one for the Tx FEC encoder, the Rx DSSS acquisition engine, or the Rx FEC decoder.

---

[17] A XC6VLX240T FPGA takes 70 Mb for full configuration. It can be configured over a 32-bit bus at 100 MHz, which takes $70 \times 10^6/(32 \times 10^8) \approx 20$ ms.

Fig. 6.10 Cognitive radio
receiver

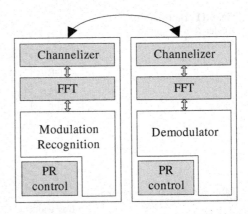

The SDR starts in transmit mode, then switches to receive mode, and then back to transmit. DSSS acquisition block typically requires a lot of resources but is not used after the spreading code is found. Once the spreading code has been acquired, those resources can be reconfigured as the FEC decoder while the Rx demodulator continues to track the received signal.

6.2.3.3 Cognitive Radio

Another example for the use of PR in SDR applications comes in the field of cognitive radio (CR, see Sect. 3.4). A CR receiver must scan the available spectrum, locate energy peaks, create a channel that attempts to match the spectral shape, perform modulation recognition, and then try to demodulate. Because the spectrum must always (or at least very often) be monitored, the FFT module is left as static, but use of the modulation recognition and demodulator are mutually exclusive, providing an opportunity to take advantage of PR. It is expected that switching between modulation recognition and demodulation will occur often as the receiver searches the available signals for the correct one. If reconfiguration could be achieved quickly enough, the FFT module could potentially be made reconfigurable as well. Partitioning for this CR receiver example can be seen in Fig. 6.10.

6.2.3.4 Hardware Acceleration

PR can be effectively applied to hardware acceleration (see Sect. 6.6). Hardware acceleration (HA) is used when some functions (e.g. DSSS acquisition, dispreading, and FEC decoding) in a GPP-based SDR cannot be executed in real time. These functions can then be offloaded to an FPGA-based accelerator.

If some latency is acceptable and enough memory is available, it is possible to implement a multi-channel receiver in software while buffering data and time-

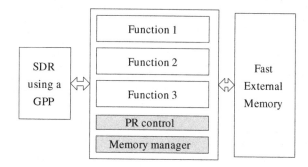

Fig. 6.11 Hardware acceleration of software-defined radio

Table 6.1 Partial Reconfiguration Interfaces

Port	Bus width	Max frequency (MHz)	μs/Frame[a]
External serial port	1 bit	100	13
External serial JTAG port	1 bit	66	20
External parallel SelectMap	8 bits	100	1.6
Internal access port (ICAP)	32 bits	100	0.4

[a] Configuration of one frame requires 41 32-bit words of configuration data

sharing the FPGA as necessary to process the data when necessary. Using an embedded microprocessor and a memory manager, portions of the FPGA could be partially reconfigured (similar to a context switch) to process one channel's data and then reconfigured to process another channel's data. Because the data can be buffered during the reconfiguration, the only limitation is the average throughput of the FPGA. Fig. 6.11 shows a block diagram of such a system.

6.2.3.5 Partial Reconfiguration in the Virtex-4

It is important to be aware of the performance and limitations of PR implemented on the target device. Because of the widespread use of Xilinx's FPGAs, the Virtex-4 family was chosen as the example FPGA.[18] The FPGA is configured by writing bits to its configuration memory. The configuration data is organized into frames that target specific areas of the FPGA through frame addresses. When using PR, the partial bitstreams contain configuration data for a whole frame if any portion of that frame is to be reconfigured.

Reconfiguration time is highly dependent upon the size and organization of the PR region(s). Four methods of configuring a Xilinx FPGA are listed in Table 6.1.

Estimated sizes and corresponding configuration times for a few common blocks are given in Table 6.2. Note that approximately 10 % overhead is incurred in setting up the addressing for PR. These values assume PR region resource

---

[18] PR has not changed significantly in the Virtex-6 and -7 devices.

Table 6.2 Configuration sizes and times for a few common blocks

Design	Frames	JTAG	ICAP (ms)
Turbo Encoder[a]	154	3.061 ms µs	0.061
Turbo Decoder[a]	4092	81.34 ms	1.6
SS Acquisition[b]	5610	111 ms	2.2
FFT [c]	4752	94.47 ms	1.9

[a] Turbo product code with block size of 16 kb
[b] Custom design capable of checking 3,000 hypotheses in parallel
[c] 8192-point FFT with 12-bit inputs

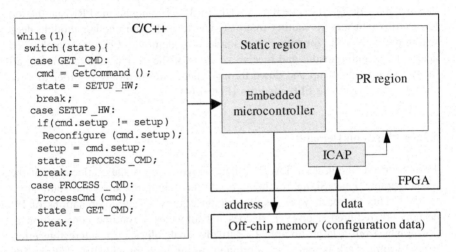

```
                          C/C++
while (1) {
  switch (state) {
   case GET_CMD:
    cmd = GetCommand ();
    state = SETUP_HW;
    break;
   case SETUP_HW:
    if (cmd.setup != setup)
     Reconfigure (cmd.setup);
    setup = cmd.setup;
    state = PROCESS_CMD;
    break;
   case PROCESS_CMD:
    ProcessCmd (cmd);
    state = GET_CMD;
    break;
```

Fig. 6.12 PR design using embedded microcontroller

utilization of 90 %. Actual sizes and times depend on design implementation, target device, region utilization, and resource location.

The simplest PR architecture relies on a microcontroller implemented in the static region of the FPGA and the ICAP interface (Fig. 6.12).

When needed, the microprocessor loads the desired configuration data from external memory and reconfigures the PR region. Reconfiguration can be triggered by an event in the FPGA itself (e.g. acquisition of a PN code, loss of signal lock, detection of an interfering signal, a set timer), or by an external interrupt. Configuration data may be loaded a priori or by the FPGA itself (for example, the FPGA is configured as a receiver downloads a new configuration).

Xilinx provides a sample code using an embedded microcontroller for partial reconfiguration. The embedded microcontroller can be easily replaced with a custom state machine that handles the loading of configuration data when the overhead of an embedded microcontroller is undesirable.

Xilinx development tools require that the top-level module contain submodules that are either static modules (SMs) or partially reconfigurable modules (PRMs). All communication (with a few exceptions for global signals such as clocks) must

Fig. 6.13 Example design
showing two PR regions

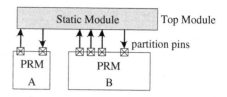

be explicitly declared using *partition pins*. An example design with two PRMs is shown in Fig. 6.13.

The required hierarchy adds a significant amount of effort when converting an existing static design into one that is ready for PR. Having all the PRMs at the top level will often require routing many signals to and from another module deep within the main static module. For example, a modem with FEC cores embedded deep within both transmit and receive portions of the design requires routing all the necessary signals up and down through the entire hierarchy in order to have partially reconfigurable FEC modules (Fig. 6.14).

6.2.3.6 Wires on Demand

The partial reconfiguration flow described above requires static allocation of the reconfigurable areas. Signal routing between reconfigurable areas and static areas is fixed. This approach works well as long as only a few combinations of PR designs are required. However, it does not allow easy runtime construction of flowgraphs from a large library of relatively small blocks. An interesting effort from Virginia Tech is aimed at improving on the tools provided by Xilinx. The 'wires-on-demand' concept is based on a single large PR area. The tool is then able to automatically place and interconnect an arbitrary number of blocks within the PR area. An example of this approach applied to SDR is described in [75]. The need for wires-on-demand is a reflection of relatively poor support for PR from Xilinx. When and if PR becomes mainstream, Xilinx will likely provide a more robust toolset that will make wires-on-demand obsolete.

6.3 Host Interface

Most SDRs can be logically separated into a block that implements a waveform and a block that decides which waveform to use. For this section, the 'decider' is known as the *host,* while the signal processing is done by the *device.* Note that the host and the device may very well be on the same chip. The host is usually a GPP, while the device can be either a GPP or hardware based (FPGA, ASIC).

In the simplest scenario, the host loads (for an FPGA-based SDR) or executes (for a GPP-based SDR) the desired waveform. If the waveform has no additional

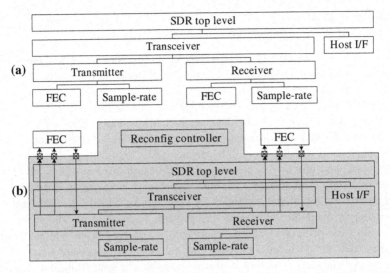

Fig. 6.14 Hierarchical firmware view of transceiver with FEC. **a** Static design. **b** Reconfigurable

parameters that must be set, the host is done. But in many cases the loaded configuration (executable) can support multiple varieties of a given waveform. For example, the data rate may be a runtime selectable parameter. Some configurations (e.g. those in Sect. 6.2.1) can have hundreds of run-time parameters. The logical interface between the host and device should be carefully considered.[19]

Interface to a GPP-based device is dictated by the SDR software infrastructure (e.g. SCA, GNU Radio, etc.). The logical interface to a hardware-based device is frequently up to the designer.

6.3.1 Memory-Mapped Interface to Hardware

The simplest logical interface paradigm for a hardware device is memory-mapped access. Each configurable parameter is given a unique address. A physical or logical bus is established between the host and the device. The bus consists of five required and two optional signals:

1. Address
2. Read_Enable
3. Read_Data
4. Write_Enable
5. Write_Data

---

[19] The author has reworked the logical interface for one complex FPGA-based SDR three times in 6 years to support growth and maintainability..

6. [Optional] Read_ready
7. [Optional] Write_acknowledge

The bus is broadcast to all the blocks in the device. Each configurable parameter is compared to the address and a parameter is set if the address matches. For example, the host would execute.

```
WriteRegister(A_SET_DATARATE, DATARATE_1Mbps)
```

To cause the device to set the data rate to 1 Mbps. Reasonable software design practice dictates that the addresses and (usually) data are given descriptive names. This interface is familiar to all hardware designers. Despite its simplicity a memory-mapped interface raises a number of design issues. Address assignment has to be carefully considered. Multiple developers working on different blocks must be able to get unique addresses for their blocks. Ideally, a centralized web-based tool is available to request and register each address. The tool is then responsible for generating source code with a mapping from address name to address value for each programming language used by the team (e.g. a VHDL package, a C header). The tool is also responsible for generating documentation with a list of registers and their function. In the simplest designs, the address is a set of sequential integers, starting from zero. However, that approach very quickly becomes impractical.

Consider a case where two independent development efforts are merged into a single design. For example, a SDR is merged with a channel emulator. Each team has its own address mapping, and the address values between the two designs are likely to be duplicated. One option is to renumber all the addresses. Unfortunately, this immediately obsoletes all printed documentation.[20] Now consider a case where multiple copies of the same design are instantiated to implement a multi-channel SDR. The addresses for all the channels are now the same and it is impossible to configure each channel for a different waveform. One solution is to assign new addresses to every instance of the design, but it makes the host software difficult to read and maintain. Instead, a hierarchical address structure should be adapted. At the lowest level of hierarchy, the addresses are sequential integers. The next level of hierarchy defines the instance number (channel). The next level of hierarchy defines the top level design (e.g. SDR, channel emulator, video processor). The address should be treated as a structure with different fields. For a hardware implementation, each field would be assigned a specific number of bits. For example, if the system is expected to have no more than four different designs and each design has no more than 16 channels, $D = 2$, $C = 4$ (Fig. 6.15).

This addressing scheme works well as long as the host accesses only one register at a time. There are (infrequent) cases where the host may need to configure a register in multiple channels or even designs *simultaneously*. For example, a command to start processing data may be issued to the receiver and an

---

[20] Hardware developers have an uncanny ability to remember four-digit hex values corresponding to addresses in the blocks they have designed and get very frustrated trying to set a register that has been renumbered..

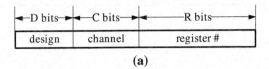

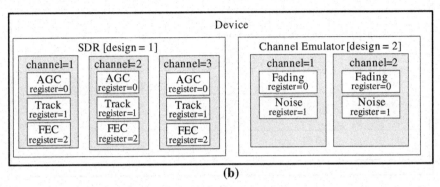

Fig. 6.15 Hierarchical memory-mapped addressing. **a** Address partitioning. **b** Hardware view

acquisition engine at the same time. One solution is to create a register in a level of hierarchy above both designs that would be written to and would in turn trigger the designs. However, the two designs may not be even co-located. A set of *broadcast* addresses can be defined (but used sparingly). One approach is to designate value of *zero* to mean *broadcast* in each level of hierarchy. A conceptual view of this addressing scheme is shown in Fig. 6.16. The address is represented by a tuple {device:D, channel:C, register:R} in Fig. 6.16. Note that the hierarchy is entirely logical and can be interpreted either as a tuple or as single value.

If interpreted as a tuple, each level of hierarchy checks the address and passes the memory map bus down[21] to the lower level of hierarchy if the address matches the mask. Looking at Channel 2 in SDR, we see that SDR level of hierarchy will pass addresses that have either 0 or 1 for the device field. 1 corresponds to SDR, while 0 is the broadcast identifier. Channel 2 level will pass addresses that have the channel field set to either 0 or 2. 2 corresponds to the channel number, while 0 is the broadcast identifier. Blocks at the lowest level of hierarchy only check if the register field matches the value(s) assigned to that block. If interpreted as a single value, the value itself is formed by an appropriate macro that combines the fields. The second approach is less elegant, but is more compatible with a hardware interface standard known as MHAL (see Sect. 7.1.3).

The reader has, no doubt, noticed that the hierarchical addressing scheme described above is very similar to the IP addressing scheme used to route packets over the Internet. Indeed, IPv6 addressing can be used for large and software-

---

[21] In hardware, the read and write signals can be gated (logically AND'ed with) by the mask. In software, the lower level functions are simply not called.

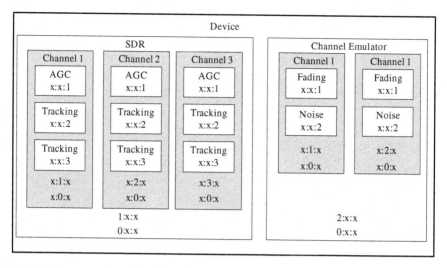

Fig. 6.16 Representative addressing scheme with hierarchy and broadcast

centric SDR designs. An immediate advantage of using IP is the ability to easily split SDR among multiple boards.[22]

6.3.1.1 Acknowledge Signals

The read ready signal is not part of the classical memory map interface.[23] However, it is quite useful for two reasons:

1. It makes detecting software and hardware bugs related to addressing much easier. For example, if no unit responds to a particular read request, the host needs to know that the read failed. A classical memory map interface would return a random value. Likewise, if more than one unit responds to a read requests, the addresses must have been misassigned and an error should be generated.
2. It allows different units to take a different amount of time to respond to a read request. For example, some units can respond immediately, while others need a few clock cycles. Once the data are ready, the unit asserts read ready.

A write acknowledge signal is also useful in detecting misassigned addresses.

---

[22] This discussion covers only the control plane. The data plane is addressed in Sect. 7.4.
[23] It is part of all modern bus interfaces such as PCI or DDR3.

Fig. 6.17 Standard message
structure for MHAL
communication

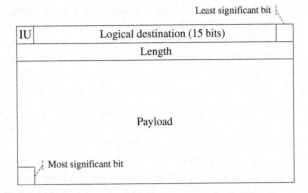

Least significant bit

IU	Logical destination (15 bits)	
	Length	

Payload

Most significant bit

6.3.2 Packet Interface

The memory map interface can be extended to support multiple data values (a *burst*) for each read/write command. In fact, most modern memory interfaces support burst read/write operations to increase throughput. A burst read/write can be thought of as a packet, with the first element interpreted as the address and the rest of the elements treated as data. It is trivial to convert from a packet interface to a classical memory map interface by simply holding the address and read/write line constant for the duration of the packet.

The MHAL API [76] defined for JTRS radios (see Sect. 7.1.3) uses the packet paradigm (Fig. 6.17). The address is referred to as the 'logical destination'. Note that, MHAL provides no guidance as to how the logical destination is to be used. It can be used to implement either a flat or hierarchical address space.

6.4 Architecture for FPGA-Based SDR

6.4.1 Configuration

The highly parallel architecture of FPGAs is ideally suited for dataflow processing, but is not well suited for control-intensive processing. Each of the reconfiguration approaches described above requires a controller to select the desired waveform. The controller may write to registers, load a complete or partial configuration file, or perform some combination of these operations. The controller is almost always implemented in software running on a GPP. This combination of a GPP and one or more FPGAs raises interesting computer architecture questions. FPGA devices available in 2011 suggest three options:

1. A dedicated GPP device is used alongside FPGA(s). Any GPP can be selected from the very wide range of available choices. The GPP has relatively modest requirements since it does no signal processing. An appropriate device belongs to the class of microcontrollers rather than microprocessors. The same GPP can also be used for user interface (if any). The memory access bandwidth requirement depends on the configuration approach:

   a. High-bandwidth access to memory is needed if partial reconfiguration is used and the waveforms must be changed quickly. Some of the lower end microcontrollers would not be suitable. As discussed above, the GPP must read the partial configuration file and write the data to the FPGA. There are two basic options for storing the configuration data: NVRAM (e.g. FLASH) and SDRAM.[24] NVRAM typically shares the memory bus with the FPGA, while SDRAM is on a separate bus. NVRAM data throughput is much lower than for SDRAM. These two considerations motivate the use of SDRAM for partial reconfiguration applications.
   b. FPGA vendors have recognized the need for fast and simple configuration and FPGA-friendly FLASH memory is available [77,78]. These devices allow the GPP to control and initiate the configuration without having to transfer the data itself. This way even a low-end microcontroller can achieve fast reconfiguration. This approach only works for full reconfiguration.
   c. Memory access is not a concern for multi-waveform configurations since only a few registers have to be set to change a waveform.

2. A full-featured microcontroller can be embedded in the FPGA by using the available logic resources.[25] This architecture raises a 'chicken vs. egg' dilemma—the microcontroller does not exist until the FPGA is programmed, and therefore cannot program the FPGA. An external configuration source is required to load the initial configuration. The embedded microcontroller can then perform waveform selection for the multi-waveform and partial reconfiguration modes. The embedded microcontroller, combined with option (1b) above, can also select a waveform in the full reconfiguration mode.

3. An FPGA and a dedicated GPP can be combined on a single device (e.g. ZYNQ from Xilinx [62]). Unlike an embedded microcontroller, the on-chip micro-controller is active upon power-up and does not require the FPGA to be pro-gramed. Other than single-chip integration, the design considerations are the same as in (1).

---

[24] Data are loaded into SDRAM prior to reconfiguration. It may be loaded from nonvolatile storage (disk, FLASH, etc.), or received over the air.

[25] A hard macro microcontroller may also be available on some FPGAs (e.g. PowerPC on Xilinx Virtex 2 and Virtex 4 FX series).

6.4.2 Data Flow

The data flow architecture is probably the most critical decision in the FPGA-based SDR development. The two major options are:

1. Datapath with flow control
2. Bus-centric

Datapath architecture is the simplest and most obvious choice. Signal processing blocks are connected in a contiguous flowgraph, closely mirroring the system-level block diagram. A block operates on the incoming samples one or more at a time and passes the output to the next block in the flowgraph. For example, in a receiver the A/D data block passes samples to the DC-offset canceling block, which, in turns, passes data to a frequency offset compensation block, etc....

The simplest implementation uses one-way handshaking—either pull (Fig. 6.18b) or push (Fig. 6.18a). In a one-way push handshaking, an upstream block indicates that data are ready to a downstream block. The downstream block *must* be able to accept and process the data. Likewise, for one-way pull handshaking a downstream block requests data from the upstream block. The upstream block *must* be able to supply data as soon as it is requested. One-way push handshaking can be used quite successfully for relatively simple receivers, while pull handshaking can be used for transmitters.

One-way handshaking fails if one or more of the blocks cannot process data as soon as it comes in, even though it can process the *average* throughput.[26] The offending block must buffer bursts of data to avoid losing it. The buffer (a FIFO) must be large enough to hold the longest expected burst. DSP blocks designed for SDR are meant to be flexible and work with a wide range of waveforms. The maximum burst size may, therefore, be unknown, or very large. It is inefficient (and inelegant) to require these 'handshaking' buffers to be sprinkled throughout the design. Classical two-way handshaking overcomes this problem. In two-way handshaking, an upstream block tells the downstream block that data are ready while a downstream block requests the data from an upstream block. Data are only transferred if both `ready` and `request` are true (Fig. 6.18c). Buffering may still be required when using two-way handshaking—one at the source and one at the sink.

Datapath architecture works well for high-speed designs where each block is operating close to its maximum throughput. However, the architecture makes it awkward to share blocks among multiple data streams.

Consider a receiver supporting two data channels at 1 and 2 Mbps. The first few blocks process A/D samples for both data streams. However, after the channelizer the two streams are separated. Assuming that all the blocks are capable of processing 3 Mbps on average, blocks after the channelizer should be shared.

---

[26] Clearly, if the block cannot handle even the average throughput the system will not work.

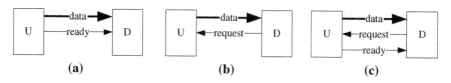

Fig. 6.18 Handshaking. **a** One-way push. **b** One-way pull. **c** Two-way

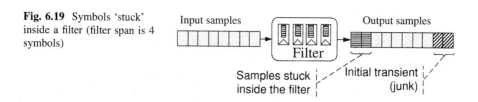

Fig. 6.19 Symbols 'stuck' inside a filter (filter span is 4 symbols)

The simple `ready/request` flow control cannot be applied anymore.[27] Furthermore, blocks that retain internal state from sample to sample (e.g. filters) must be designed to store states for both channels and automatically switch between them.[28]

The fixed flowgraph also limits flexibility. Consider the example above, with different waveforms used for 1 Mbps and 2 Mbps channels. For one channel, the frequency tracking block should come before the matched filter, while for the second channel it should come after the matched filter.[29] The control and routing logic to switch the position of a block in a flowgraph is quite messy.

Datapath architecture works well for streaming waveforms (i.e. symbols are transmitted continuously). However, it is quite inconvenient for packet-based waveforms. For a packet-based waveform, the transmitter can send out a (relatively) short message and the receiver must be able to demodulate short messages. The problem with a datapath architecture is data in the pipeline. Consider a simple FIR filter used for pulse shaping. The filter has some memory (4–8 symbols is typical), meaning that the last few symbols of a message are still inside the filter after the last symbol of the message has been input to the filter (Fig. 6.19). The filter memory translates into a small latency for streaming data. After the message is done, the 'stuck' symbols get pushed out by the next message. However, if messages are not contiguous, the stuck symbols do not get transmitted until the *next* message is sent. The problem can be overcome by intentionally inserting some padding zero symbols after each message. Of course, this requires the datapath to be aware of the message boundaries and whether there is a gap between messages.

---

[27] It is quite possible to design a module to switch the flow control logic between the two channels, but based on the author's experience the code quickly becomes unreadable..

[28] This is also quite possible, but leads to lower performance, and requires significantly more development effort.

[29] This somewhat artificial example may be motivated by different expected frequency dynamics for the two channels.

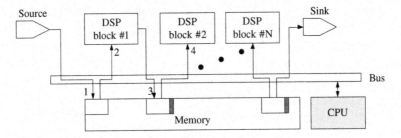

Fig. 6.20 Basic bus-centric architecture

A bus-centric architecture replaces point-to-point connections between blocks with a shared bus. This architecture is also known as processor-centric or system-on-a-chip. Each of the DSP blocks is treated as a peripheral to a central processing unit. The CPU can be implemented on a low-end embedded GPP or even a dedicated state machine. A simple example is shown in Fig. 6.20. A block of received samples is buffered in memory (Fig. 6.20:1). The CPU then transfers the block of samples to the first DSP block (Fig. 6.20:2). The output of the DSP block is again buffered in memory.[30] The internal state of the DSP block may also be saved to memory to allow seamless processing of multiple data streams (shaded block in Fig. 6.20). Once the entire block has been processed, the CPU transfers the output block to the second DSP block (Fig. 6.20:4). The process continues until all the processing has been applied to the data.

A major problem with the architecture and dataflow described above is very low performance and throughput. Indeed, all but one of the blocks are idle at any given time. The operation is very similar to executing serial (not multi-threaded) software. Although it makes poor use of FPGA resources, this architecture works if high-throughput is not required.[31]

The key to an efficient bus-centric architecture is to offload the actual movement of data between blocks from the CPU to a sophisticated DMA engine. A detailed discussion of DMA engine and bus architectures is beyond the scope of this book. The CPU is only responsible for keeping track of available memory buffers and scheduling data transfers. Actual transfers are handled by the DMA engine. A multi-channel DMA engine can simultaneously handle transfers to and from multiple blocks. Extending the example in Fig. 6.20: While block #1 is processing the first block of data from the source (Fig. 6.21:A), the source is saving data into a new buffer (Fig. 6.21:B). While block #2 is processing the output of block #1 (Fig. 6.21:C), block #1 is processing data from Fig. 6.21:B.

---

[30] In-place operations overwrite the input memory. Alternatively, the output is placed in a different part of memory.

[31] An FPGA is not the best device to implement a SDR with low-throughput requirements and a GPP or a DSP may be more appropriate.

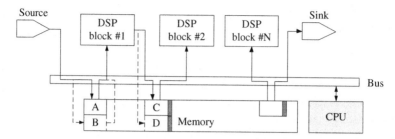

Fig. 6.21 Pipelined bus-centric architecture. *Dashed lines* indicate data flow for even numbered block periods, while *solid lines* are for odd numbered

All the blocks are fully utilized in the pipelined bus-centric architecture, assuming two constraints are met:

1. Bus throughput is sufficient to sustain all the simultaneous data transfers. For example, a system with 10 blocks and a nominal sample rate of 5 MSPS must have a bus capable of sustained throughput of $2 \times 10 \times 5 = 100$ MSPS. The conventional shared bus topology shown in Fig. 6.20 does not scale with the number of blocks and can quickly become a bottleneck. Memory throughput is also a potential limiting factor, since it has the same requirements as the bus throughput.
2. All of the blocks process data at the same rate. It is easy to see that the overall throughput is determined by the throughput of the slowest block. Fortunately, the bus-centric architecture allows makes it easy to replicate slow blocks to achieve higher throughput.

6.4.3 Advanced Bus Architectures

At least three different bus topologies can be applied to alleviate the bus and memory throughput constraint identified above. However, before considering the alternatives let us consider when the constraint becomes an issue. The physical layer of a typical radio consists of 15–30 discrete DSP blocks. Assume each block processes complex baseband[32] data sampled at X MHz with 16 bit precision. Bus throughput, B, is then

$$30 \text{ blocks} \times 2 \text{ samples/complex pair} \times X \text{ MHz} \times 2 \text{ read/write} = 240X \text{ MBps}$$

A modern FPGA can support a 256 bit data bus operating at 200 MHz, giving a throughput of 6.6 GBps. Assuming, for simplicity, 100 % bus utilization, that bus can support $X = 6.6\ GBps/240\ MBps \sim 25\ MHz$. A complex sample rate of 25 MHz supports signal bandwidth up to 10–20 MHz. This bandwidth covers most

---

[32] Complex baseband sample-rate data dominates overall throughput.

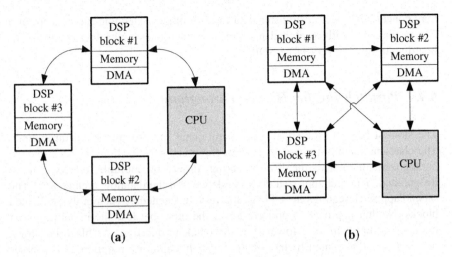

Fig. 6.22 Higher performance bus architectures. **a** Ring. **b** Full mesh

of the commercial standards in use today and may be sufficient for a wide range of SDR applications. However, a higher performance bus is required for wideband radios (e.g. 100 MHz bandwidth). A higher performance bus is best served by a distributed memory architecture. Instead of centralizing memory and DMA, each block has its own.

1. If the dataflow is expected to be mostly between successive blocks, a ring topology is appropriate (Fig. 6.22a). Note that, this architecture looks quite similar to a datapath and therefore suffers from the same problems. In particular, changing the order of blocks is no longer simple.
2. The most flexible solution is a full or partial mesh (Fig. 6.22b). The CPU is still responsible for configuring the DMA engines in each block, but all data transfer is now point-to-point. The main disadvantage is that the number of connections grows as a *factorial* of the number of blocks. Routing all these connections becomes impractical for even a few blocks. A partial mesh prunes many of the connections, making the hardware simpler at the cost of potentially reduced flexibility.

The reader may wonder why the FPGA fabric itself cannot be reconfigured to implement just the desired connectivity. After all, an FPGA is essentially a sea of logic gates with programmable interconnections. Unfortunately, the programmable interconnection is mostly local, used to connect adjacent logic gates. A major change in routing (e.g. exchanging the location of two blocks) requires a new placement of the logic gates.

A discussion of bus architectures would be incomplete without mentioning the 'network-on-a-chip' approach (see Sect. 5.4.1 for an example). The mesh architecture can be implemented by routing each bus to a port of a switch. This architecture does not scale well because the switch quickly becomes too large. Alternatively, a hierarchical mesh structure can be formed by grouping blocks that may conceivably need to transfer data to each other. Extending the concept of a bus to that of a fully switched network also facilitates multi-device and multi-board solutions for very large SDRs.

6.4.4 Parallelizing for Higher Throughput

The slowest block in the signal processing chain sets the maximum throughput. The classical solution to increasing throughput is parallelization. A block may be parallelized internally, thereby appearing faster, or multiple blocks can be instantiated. The first approach does not depend on the dataflow architecture. The second approach requires the CPU to appropriately schedule each of the replicated blocks. Neither approach is preferred over the other because the implementation burden is simply shifted. However, if the block in question is a black box (or a licensed core that cannot be modified), only replication is an option. Four classes of blocks must be considered, with different approaches to parallelization for each:

1. Memoryless blocks process data one sample at a time and possess no internal state. Examples include: upconverter, fixed DC offset canceler. These blocks are simply replicated.
2. Some blocks process a fixed set of samples at a time and then reset to the initial state. Examples include: block FEC codes (Reed-Solomon, Turbo, LDPC), terminated convolutional codes, block interleavers, fast Fourier transforms. These blocks are simply replicated and the CPU makes sure that one or more complete sets are processed at once.
3. Finite memory blocks process data one sample at a time, but possess an internal state. The output depends on the current *and* some of the past input. Examples include: FIR filters, resamplers. These blocks are more difficult to parallelize because the internal state becomes invalid for discontinuous input. One solution is to overlap data processed by each instance of the block. The overlap, N, must be long enough to account for the internal state. The first N outputs from each instance are then discarded. As long as N is small relative to the number of samples in a chunk, the overhead is negligible
4. Blocks with infinite memory have internal feedback. The output depends on *all* of the past input. Examples include: IIR filters, phase and symbol tracking loops, automatic gain control loops. These blocks are by far the hardest to parallelize and must be considered on a case-by-case basis. If at all possible, these blocks should be parallelized internally to hide the complexity from the CPU. For example, look-ahead computation can be used to speedup IIR filters (see Chap. 10 in [79]). Feedback loops for tracking and adaptation may be parallelized if the loop bandwidth is small relative to the duration of the chunk of samples processed at once.

The technique of parallelization is obviously not specific to the design of SDR. However, it does come up a lot since the same SDR hardware platform may be

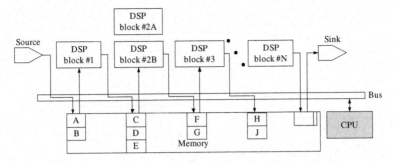

Fig. 6.23 Duplicating a slow block to increase throughput

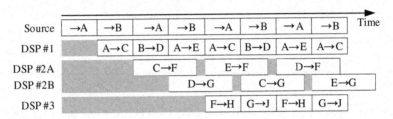

Fig. 6.24 Timing diagram of a pipelined and parallelized architecture

used to support signal bandwidth from <1 MHz to 100s of MHz. It is often desirable to reuse the same signal processing blocks for all the applications. Consider an example, where the second DSP block in Fig. 6.21 takes about twice as much time to process a chunk of data as all the other blocks. Assume that block-level parallelization is feasible because the block belongs to the first three classes described above. Another instance of DSP block #2 and an associated memory buffer is added to the original block diagram in Fig. 6.21, resulting in Fig. 6.23.

Operation of the pipelined and parallelized architecture is best explained via a timing diagram as shown in Fig. 6.24. Each line indicates what buffers are used by a given functional block. Note that, blocks 2A and 2B each take about 1¾ time slots to complete processing of data that takes all other blocks 1 time slot. These two blocks are idle about ¼ of the time, but allow the rest of the pipeline to operate at full efficiency. Buffer management performed by the CPU is somewhat more complicated.

6.5 Hybrid and Multi-FPGA Architectures

One of the main disadvantages of FPGAs relative to DSPs and especially ASICs is very high power consumption. Power consumption when the FPGA is not processing any data (idle) is especially bad relative to other devices. Indeed, static

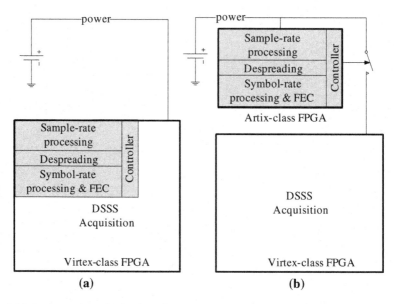

Fig. 6.25 Implementing a SDR with DSSS support on an FPGA. **a** One device **b** Low power

power consumption due to leakage can easily account for over half of the total power [80]. Not surprisingly, static power consumption for the high end, high-performance FPGAs is worse than for the lower end versions. FPGA vendors are well aware of this problem and are actively working to address it [80, 81]. However, for the foreseeable future, high-end FPGAs will continue to draw a lot of power when idle. Power consumption motivates a multi-FPGA or a hybrid FPGA/DSP architecture.

Consider a SDR that supports a wide range of waveforms, including some that are spread. As mentioned previously, DSSS acquisition and FEC decoding are typically the most computationally intensive blocks in a SDR. This particular radio is used for relatively low rate communications (<1 Mbps). All of the signal processing, except for DSSS acquisition, can be easily implemented on a low-power FPGA such as Spartan/Artix/Cyclone or a DSP. Acquisition alone requires 10 × more computational resources than the rest of the radio. This is not at all unreasonable if the spreading gain (see Sect. A.4.8) is large and the spreading code is either nonrepeating or has a long period. The floorplan for a single-FPGA solution would look like Fig. 6.25a.

The code offset for a streaming (vs. packet-based) DSSS waveform[33] has to be acquired once at the beginning of communication and is then tracked by the

---

[33] GPS is a classical example of a streaming DSSS waveform. WiFi 802.11a is a classical example of a packet-based DSSS waveform.

symbol-rate processing in the receiver. Thus, the acquisition engine is idle most of the time while SDR is receiving a streaming DSSS waveform. However, the transistors dedicated to implementing the engine continue to draw standby power. Assuming standby power is about half of the total power, the idle acquisition engine can use $10 \times \frac{1}{2} = 5$ times more power than the active part of the SDR. This is clearly unacceptable for a low-power SDR. An obvious solution is shown in Fig. 6.25b. The infrequently used block is moved to a separate FPGA, while the rest of the SDR is implemented on a small and low-power FPGA (or a DSP). The power supply to the large FPGA is enabled[34] by the controller on the small FPGA whenever acquisition is needed, and is disabled the rest of the time. This approach is known as a 'power island' and is widely used on large ASICs. It is only a matter of time until FPGAs have user-controlled power islands, but until then a two-FPGA solution can result in dramatic power reduction.

An unusual radio used for a sensor in a wireless sensor net provides another example. A sensor measures concentration of a particular chemical and transmits the results on a regular basis. Very infrequently, the sensor is scheduled to receive an update (perhaps change the chemical to monitor). A demodulator for a given waveform is almost always significantly more complex than the modulator. It would be quite inefficient to always have the receiver loaded in the FPGA, wasting standby power. Instead, the large FPGA is turned on and configured with the receiver logic only when the node is scheduled to receive an update.

The general concept of using a low-power device most of the time but having a lot of signal processing resources 'in reserve' is not unique to FPGAs. However, it is not as effective when applied to DSPs or GPPs because these devices already support power islands on-chip. Leakage power is less of a problem and power consumption can be scaled by changing the operating clock frequency or putting the device in various low-power modes.

6.6 Hardware Acceleration

A hardware-accelerated architecture follows naturally out of the bus-centric FPGA architecture. A GPP is the best platform for SDR development if flexibility and rapid development are more important than size, weight, and power. This is the case for many research SDRs and for SDRs in ground stations. Unfortunately, even

---

[34] At least three solutions should be considered for the power 'switch': (1) low dropout regulator (LDO), (2) dedicated DC/DC converter or a separate channel on a multi-channel DC/DC, and (3) a power FET. LDOs are generally not recommended since FPGAs do not require a particularly clean main digital supply, making LDO unnecessary. Further, even a small voltage drop across the LDO results in significant power dissipation since the current is very high. A dedicated DC/DC is a good choice if board space and cost are not major drivers. However, a low-power device is typically also a small device and may not have the additional space required for a DC/DC. A power FET acts as an almost ideal switch with a resistance of about 2 mΩ.

the fastest GPPs available today cannot process wideband and complex waveforms in real time. Analysis of the throughput bottleneck in a typical radio receiver on a GPP reveals one or two functional blocks that consume the bulk of computational time. In particular, GPPs are not well suited for two common receiver functions: DSSS acquisition and FEC decoding. These functions require many identical operations on low-precision data. A GPP is optimally utilized when the data precision is equal to the internal datapath width—usually 64 bits. Data precision for DSSS acquisition is typically only 2–4 bits, while soft decisions used for FEC decoding are typically less than 5 bits. FPGAs, on the other hand, excel at this type of processing. An obvious solution is to offload a minimum[35] set of operations from the GPP to an FPGA that would allow the combined system to achieve real-time throughput.

The concept of HA has been around since the 1980s, when the early fixed-point processors (e.g. 80386DX) used coprocessors (80387) to compute floating-point arithmetic operations. Coprocessors have become a ubiquitous part of today's computers in the form of graphics processing units (i.e., graphics cards). HA does not always mean having a separate device—HA blocks are routinely integrated on the same die as the GPP. For example, DSP chips from TI include multiple HA blocks to accelerate error correction decoding, while GPPs from Intel now include a HA block to accelerate cryptography. While the application of ASIC HA has proven itself valuable over the years, the value of flexible or reconfigurable HA has been uncertain. As little as one decade ago, work was done that concluded that the technology at that time did not provide enough gains to warrant the efforts of HA [82]. While the idea of accelerating software through the use of specialized hardware is not new, the advancement of today's HA boards along with increasing transfer speeds between the main processor and the acceleration hardware has made general HA an attractive alternative to the current software-only solutions. HA is closely related to the concept of reconfigurable computing [83].

Despite compelling performance improvement offered by HA, it remains architecture of 'last resort'. Software developers will only consider HA if the alternative is a pure FPGA-based architecture. The reluctance is understandable since a different skill set is now required in the team to develop the HA logic.

6.6.1 Software Considerations

The goal of a HA development is to closely resemble an all-software development. The HA interface API must be designed to be transparent to the software developer. Indeed, as GPP processing power scales with Moore's law, accelerated functions may be replaced with their software equivalents without requiring any

---

[35] Even if some functions are more optimally implemented on an FPGA, they should remain on the GPP as long as the throughput requirements are met.

changes to the calling function. The decision about which functions should be moved to HA is typically obvious—FEC decoding and DSSS acquisition are the first candidates. Other functions are not as clear—DSSS despreading, FFT operations for OFDM are all computationally intensive but can be done quite efficiently on a GPP. The combination of cache hits and SIMD instructions makes it difficult to estimate the performance of a given function on a modern GPP. Simply counting the number of arithmetic operations is not sufficient. Classical software engineering doctrine teaches "don't optimize unless you have to." Thus, HA is applied once an all-software SDR fails to achieve the require throughput. Optimization then begins by first profiling the SDR application using a standard tool such as Intel Vtune [84] or gprof. Profiling should identify one or a few functions that take up most of the computation time. Note that if computation time is evenly distributed across many functions, the SDR application is not likely to benefit from HA. The maximum potential performance increase is given by Amdahl's law:

$$T = \frac{1}{(1 - \alpha) + \frac{\alpha}{s}}$$

where α is the fraction of the execution time of the original software that will be accelerated, s is the speedup obtained for that portion of the software, and T is the overall speedup. For example, if a single function takes half of the computation time and that function can be accelerated to take zero time (i.e. $s = \infty$), the overall speedup is no more than a factor of 2.

When HA is finally embraced, a number of tradeoffs must be made:

1. Choice of hardware.

   a. The first choice is always a SPU, specifically a GPGPU (see Sect. 5.4). GPGPU development flow is relatively similar to GPP development and uses the 'C' or C ++ language familiar to GPP developers. Further, GPGPU board and tool suite are relatively inexpensive.

   b. FPGAs are used for HA if a GPGPU solution does not provide sufficient throughput [85], or requires too much power, or if FPGA development skills are available to the team.

2. Physical interface. Throughput and latency between the GPP and HA are often more important than the raw processing power of the HA. Vendors attempt to differentiate themselves by offering HA solutions for every available GPP interface, as shown in Fig. 6.26a.

   a. PCIe is the most common and widely available choice. A 16-lane PCIe 3.0 board offers a theoretical bidirectional throughput of 16 GB/s. Latency is below 250 ns. This throughput far exceeds the computational capabilities of today's processors—i.e. the GPP cannot process data fast enough to fill the link. A representative FPGA accelerator board is available from Nallatech [86], while a representative GPGPU board is available from Nvidia [87]. PCIe is suitable for the vast majority of SDR implementations with HA.

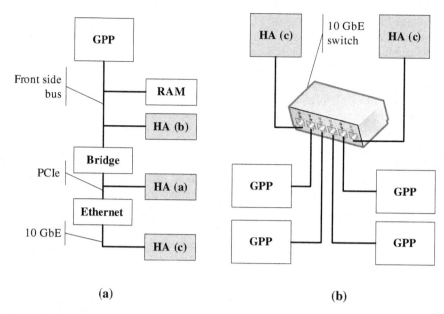

Fig. 6.26 Hardware acceleration options. **a** Direct attach **b** Switch

A high-end GPP motherboard provides up to four 16x PCIe slots [88], and external PCIe expansion solutions are available [89]. PCIe is the only option available for GPU acceleration.

b. Direct access to the processor front size bus allows exotic HA boards to achieve higher throughput, and more importantly, significantly lower latency than the PCIe bus. Representative boards are available from Nallatech for Intel GPP [90] and from XtremeData for AMD GPP [91].[36] These solutions are targeted for the scientific computing market.

c. Network-attached HAs are appropriate when one or more accelerators must be shared between many GPPs (Figure 6.26b). 10 GbE provides throughput just under 1 GB/s and latency of just under 5 μs. Note that these metrics are both about one order of magnitude worse than a PCIe connection.[37]

The effect of throughput and latency on the performance of a HA SDR is not straightforward, but depends on the specifics of the waveform. Clearly, throughput must be sufficient to support real-time transfer of data from the GPP to HA and back. For example:

---

[36] This product is apparently no longer supported. The market for front-side-bus accelerators is very limited and the benefits over PCIe accelerators are small.

[37] Lower end solutions using 1 GbE links are possible, but are not recommended since the price/performance of the 10 GbE links is significantly better.

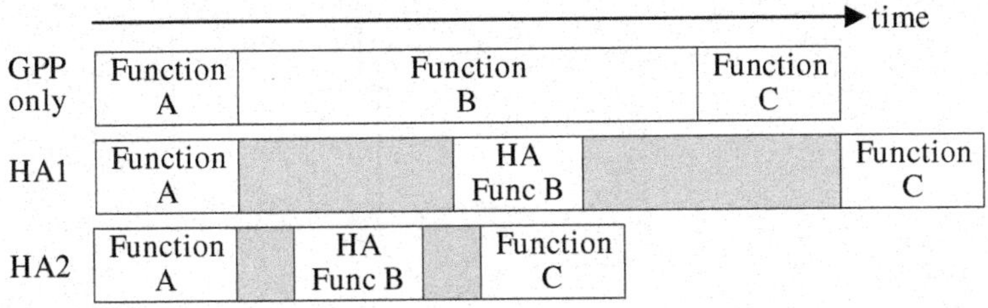

Fig. 6.27 Conceptual idea of how data transfer latency affects the benefits of hardware acceleration (*shaded blocks* denote data transfer)

A SDR receiver must process a signal sampled at 1 MHz with 8 bit precision. HA is used to implement the first frequency conversion. A minimum of 8 Mbps bidirectional throughput is required to send the raw samples to HA and get them back to the GPP. However, this throughput may be much too low in a practical application, as discussed below.

The fundamental hurdle of ensuring efficient HA is guaranteeing that the time required to transfer the data to and from the acceleration hardware is less than the time required to simply process the data on the GPP. Figure 6.27 shows a simplified example of using HA to speed up a generic *Function B*. The long transfer times incurred by HA1 make the total latency longer (lower throughput than simply using a GPP), whereas the short transfer times associated with HA2 make the total latency shorter (higher throughput). While processing data on specialized hardware is significantly faster than on a GPP, the overhead involved in transferring the data is the key metric in determining the potential performance improvement.

Direct Memory Access (DMA) allows the GPP to continue processing during data transfers. DMA is only possible if HA has direct access to the GPP memory through the PCIe or front side bus (Fig. 6.26a). Network-attached HA (Fig. 6.26b) requires GPP to handle the network packets[38] and the subsequent discussion does not apply. For DMA transfers to the hardware, the GPP is required to move the data into a special area of locked memory. The GPP can then inform that hardware that the data are ready and proceed to other processing. At that point, it is up to the hardware to retrieve the data directly from the memory without the involvement of the GPP. When the hardware has finished processing the data, it can place it directly back in the locked memory. The GPP can be informed that the processed data is available through the use of interrupts. A scenario is illustrated in Fig. 6.29 showing hardware acceleration of three different functions using DMA transfers. The GPP quickly sets up three DMA transfers and continues processing other data, while the hardware controller transfers the data.

---

[38] Technically, DMA transfers are still taking place between the GPP and the network card.

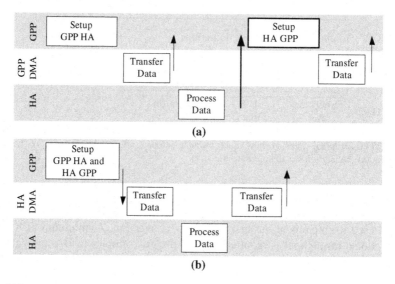

Fig. 6.28 Differences between **a** slave-only HA and **b** bus master HA (*arrows* indicate interrupts)

A device on the bus with its own DMA engine can act as the bus master (vs. slave-only) and independently initiate data transfer to and from system memory.[39] All modern GPPs have a DMA engine and are therefore bus masters. At first glance, it appears that one DMA engine on a bus is sufficient and a HA does not need one. In fact, the simplest HA implemented in a FPGA may forgo a DMA engine. All GPU HAs do have a full-featured DMA engine. A dedicated DMA engine on HA reduces the load on the GPP by eliminating one interrupt and consolidating transfer setup (Fig. 6.28). Further, the GPP DMA is used for software processing, and DMA resources (channels) may not be available at all times.

After each DMA transfer completes,[40] the HAs start working and interrupt the GPP when the data has been processed. The GPP can then initiate the DMA read to retrieve the data from the hardware. It is important to note that for the example shown in Fig. 6.29, the bus is occupied most of the time indicating that data transfer limits overall throughput. The HA resources are idle approximately 50 % of the time due to the data transfer bottleneck.

Modern bus architectures allow simultaneous writing and reading of data. A sufficiently sophisticated DMA engine can take advantage of this feature and read output of one HA while writing data to another one.

---

[39] Incidentally, DMA can also be used to transfer data between multiple HA boards in the same system.

[40] A large class of HA can start processing data before the DMA completes. For example, a filter.

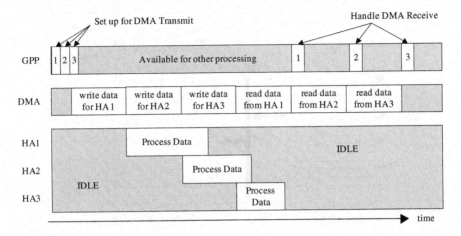

Fig. 6.29 Three-accelerator transfer scenario using DMA

6.6.2 Multiple HA and Resource Sharing

Consider a SDR processing multiple data streams in parallel. All the streams must share a single HA, resulting in potential contention. The simplest case involves two streams requesting access to the same HA. The HA API is responsible for maintaining appropriate locks (semaphores) to provide exclusive access to shared hardware. A supervising software layer needs to be developed to implement a sharing scheme (Fig. 6.30). Such a scheme may be very simple (round-robin, first-request, random-grant, etc.) or rather complex with different priority levels for each stream.

Another layer of complexity is introduced if multiple HA functions are provided, but only a subset (e.g. one) can be available at any given time. Some time is required to switch among different HA functions. This time can be very short for a GPU HA, or quite long for an FPGA HA that requires loading a new configuration. The HA supervisor must now take into account not just the immediate request(s), but the entire request queue. For example, if streams 1 and 3 request HA function A, and stream 2 requests HA function B, it is more efficient[41] to service streams 1 and 3 followed by 2 rather than servicing them in order and incurring two reconfiguration delays (Fig. 6.31). FPGA-based HA introduce another degree of freedom if multiple but not all functions can be supported simultaneously in a given configuration. For example, stream 1 requests function A, stream 2 requests function B, stream 3 requests function C. Functions A and C fit into an FPGA

---

[41] Efficiency may have to be sacrificed if the processes cannot tolerate additional latency..

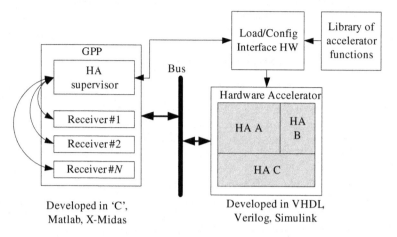

Fig. 6.30 Hardware accelerated architecture detail

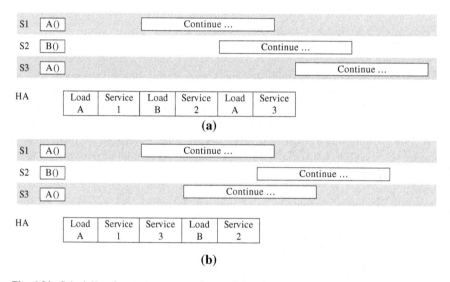

Fig. 6.31 Scheduling for resource contention. **a** Suboptimal. **b** Optimal

simultaneously, but B requires the entire FPGA.[42] The supervisor should, therefore, service streams 1 and 3 first, followed by stream 2.

Scheduling HA becomes significantly more difficult for a massively multichannel system. A cellular basestation is a representative example, with one SDR

---

[42] A plausible scenario is: functions A and C are FEC decoders, while function B is a DSSS acquisition. B takes a lot more FPGA resources and cannot share the FPGA with any other function.

Fig. 6.32 Allocating
reconfigurable regions to HA
functions

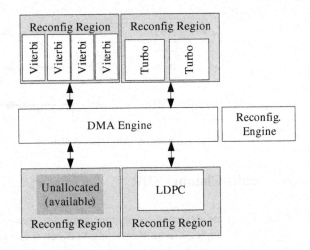

supporting dozens or even hundreds of users. Consider the case with only one HA function (e.g. turbo decoding) shared among 11 streams. Let the accelerated processing time be 10 times shorter than doing the same function in software. If the streams are asynchronous to each other, there will be a time when all streams request the HA function at the same time. The supervisor will queue up all 11 users. The 11th user will have to wait *10 × processing time* to gain access to the HA. In that case, it is better off simply executing the function in software. A robust supervisor should detect this condition and execute the function not using HA.

Partial reconfiguration (PR, see Sect. 6.2.3) is a natural fit for FPGA-based HA. In fact, it is the only truly compelling case for the user of partial reconfiguration that the author has come across. Modern FPGAs are large enough to simultaneously host multiple HA functions.[43] The desired mix of functions depends on the waveform(s) the SDR is processing. For example, a cellular basestation may have a mix of 2, 3, and 4G users. 2G waveforms use a Viterbi decoder, 3G use a Turbo decoder, while 4G use an LDPC decoder. For every new user, the HA supervisor loads an appropriate function into unallocated FPGA space. Unused functions are overwritten as needed. The key advantage of PR is that all the functions continue operating while a new one is loaded. Existing data streams are not interrupted as new ones are added or removed. As discussed in Sect. 6.2.3.5, partial reconfiguration requires the developer to define fixed-size regions with fixed interconnect to the static part of the FPGA. These two constraints may result in suboptimal utilization of the FPGA resources since the region size must be large enough to accommodate the largest HA function. Multiple smaller HA functions can be implemented inside a given region as shown in Fig. 6.32.

---

[43] For example, a Virtex-6 LX240 device can hold 30 3GPP Turbo Decoders, or 60 Viterbi Decoders, or 4 802.16e LDPC decoders.

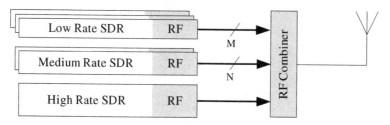

Fig. 6.33 Simple architecture for multi-channel SDR

6.7 Multi-Channel SDR

Some SDRs are designed to support many simultaneous users sharing the same band. For example, a radio translating between incompatible radios of police and the National Guard has to support dozens of simultaneous users and allow any user to talk to any other user. Each user may use a different waveform with a very wide range of data ranges (e.g. from 1 Kbps to 100 Mbps). Most users need a few Kbps for voice; others need real-time video feeds; and the command and control center can easily require 100 Mbps for satellite up/downlink. It is difficult to predict what mix of different user types and waveforms will be required at any time. This section discusses three architectures for implementing such a SDR (for conciseness only the transmitter is discussed):

1. A bank of individual SDRs connected to a shared antenna
2. Individual SDRs sharing a common RFFE and digitizer
3. An integrated multi-channel radio with a shared pool of DSP resources and a common RFFE and digitizer.

The simplest architecture for such a multi-channel SDR consists of a bank of individual SDRs, each supporting one or more channels as shown in Fig. 6.33. Each SDR has a dedicated RF front end and requires a dedicated link to the antenna(s). An RF combiner adds all the signals together before transmission. This architecture is used in most SDRs of this class deployed in 2011.

The advantages are:

- More radios can be quickly and easily added by simply connecting to unused ports on the combiner.
- Minimal coordination is required between vendors of different radios.
- Each radio can operate independently if needed and does not rely on any existing infrastructure.

The disadvantages are:

- Extensive cabling is required. RF cables take up space. Since each cable carriers the output of the radio's power amplifier, it must be able to handle the power.

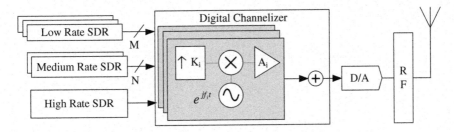

Fig. 6.34 Distributed DSP/RF architecture for multi-channel SDR

- Signal power decreases for longer cable runs, leading to reduced output power or reduced SNR (especially significant for receivers). The combiner itself also decreases power, and the insertion loss increases with a larger number of ports.
- Lack of coordination between radios can lead to radios interfering with each other. For example, two radios can be tuned to the same carrier. These errors are difficult to identify and diagnose in a high paced, high-stress environment of disaster recovery.
- Not all of the radios can be used at the same time. For example, a band can support either 10 low-rate users or 2 medium-rate users. Ten low-rate SDRs are required in case all the users are low-rate. However, if there's one medium rate user, 5 of the low-rate SDRs must be idle.[44]

A recent trend in SDR development, known as Digital IF, is leading to decoupling of DSP from mixed signal and RF. A robust standard, VITA49 (see Sect. 7.4.1), was developed to facilitate the transition from tightly coupled RF to distributed RF. In practical terms, this means that the DAC and RF upconverter need not be co-located with the DSP. A distributed DSP/RF architecture for a multi-channel SDR is illustrated in Fig. 6.34. Each SDR outputs a digital data stream,[45] preferably in VITA49 format. The data stream consists of baseband or low IF samples representing the waveform, and context packets that define the carrier frequency and transmit power. Data streams from all SDRs are combined in a digitial channelizer. Conceptually,[46] the channelizer interpolates each signal to a common sample rate required to cover the entire band. Each signal is then digitally frequency shifted to the specified carrier frequency, f_i, and scaled to the specified power level, A_i. The sum of all signals is passed to a DAC and RF front end.

The advantages are:

---

[44] Of course, a medium rate SDR may be reconfigurable to support 5 low-rate channels. However, for the purpose of this example, we assume that waveforms from one class cannot be efficiently implemented on a radio designed for a differnt class.

[45] The data are usually, but not necessarily, transmitted over a standard Ethernet/IP network.

[46] Efficient channelizer implementation does **not** have a separate interpolator and frequency mixer for each signal. Well understood techniques such as filter banks or FFT are used to reduce hardware complexity.

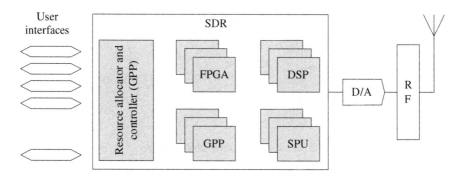

Fig. 6.35 Integrated multi-channel SDR

- Simpler cabling infrastructure if standard Ethernet networking is used. A network switch can combine one or more radios and only a single cable is run to the channelizer. Once the cable is in place, adding one more radio is as simple as plugging it into an open switch port.
- Cabling loss is eliminated since digital data does not degrade going over the network.[47]
- Each individual SDR is lower cost since an RF interface is not needed.
- The channelizer can detect and report invalid configurations (e.g. two radios using the same carrier frequency, or overlapping their signals).

The disadvantages are:

- A wideband channelizer and DAC are required. Technology in 2011 limits the bandwidth of a DAC to about 1 GHz. The DAC must provide sufficient dynamic range to allow for amplitude variation between the smallest signal and the maximum sum of all other signals.
- Some infrastructure (i.e. channelizer and RF) are required before any of the radios can be used.
- Digital IF introduces a small latency between the DSP and RF.
- Some of the radios are always unused as the mix of users changes.
- The channelizer and DAC introduce a single point of failure—if it goes, none of the radios can operate.
- A wideband amplifier must provide enough power for all the users.

The most flexible SDR architecture combines all the individual SDRs into a single radio, as shown in Fig. 6.35. A central resource allocator and controller program allocates signal processing resources to each user/waveform. The pool of resources is shared among all users and typically includes a mix of FPGAs and

---

[47] In practice, networks are not ideal and packets can be lost. Digital IF (VITA49) is always sent over an unreliable transport protocol such as UDP. Reliable protocols such as TCP introduce too much latency if lost packets are retransmitted from the source. Packet-level error correction techniques can be used to mitigate the packet loss problem [115].

GPPs. The controller configures FPGA bit loads and GPP software loads in real time, as the user mix changes. The controller also configures data flow between the DSP resources and the DAC. Sharing the resources among many users allows the entire radio to use fewer resources than separate radios. Note that this architecture is a large-scale version of the one described in Sect. 6.6.2. Some of the DSP resources are likely allocated to implement a digital channelizer like the one described above.

The advantages are:

- Efficient use of DSP resources. The peak usage is always smaller than the sum of all supported users since not all users can be active simultaneously. The savings can be quite dramatic is one user requires a large burst of resources (e.g. to acquire a highly spread, low power signal).
- Centralized control ensures that users cannot interfere with each other.
- Reduced cost to develop new waveforms since the developer can leverage a lot of existing infrastructure, both hardware, and firmware.

The disadvantages are:

- One organization must bear the cost of developing and maintaining a large and complex SDR. In other architectures, each organization is responsible for its own radios.
- Development of new waveforms for new users can be complicated since the knowledge base required to work with the complex system may be limited outside the developer of the system.
- The resource allocator and controller is a very sophisticated piece of software and is therefore time-consuming and expensive to develop.
- Centralized processing introduces a single point of failure.

Chapter 7
SDR Standardization

SDR hardware has matured to the point where wide deployment is called for. However, many potential users are concerned that lack of standards makes adopting a particular hardware platform risky since software and firmware developed for that platform may not be portable to other hardware. SDR standards must address three different parts of a waveform life cycle:

1. Describe the waveform. A waveform description should be entirely hardware independent but contain all the necessary information to implement the waveform on appropriate hardware.
2. Implement the waveform. Waveform implementation should be done in a relatively hardware-independent environment to facilitate portability to different hardware platforms.
3. Control the waveform. Once the waveform is operating, its features may be modified at runtime.

Standardization of SDR has been progressing for many years. The JTRS/SCA standard developed by the US Army [76] and STRS standard developed by NASA [92] define robust and powerful infrastructures for flexible radios. These standards are described in detail in the next two sections. There is no accepted standard for describing the *intent* rather than the *implementation* of a waveform for SDR. One potential solution is described in Sect. 7.3. Finally, Sect. 7.4 covers low-level standards for data exchange between SDR components.

7.1 Software Communications Architecture and JTRS

The US military uses and maintains dozens of different radios. Devices built to military specs last much longer than their commercial equivalents, and old devices remain in use as new versions are procured. Further, each branch of the Army has different requirements for communications (e.g., a Delta Force operative has very different needs than a tank brigade commander), and thus procures a different

E. Grayver, *Implementing Software Defined Radio*,
DOI: 10.1007/978-1-4419-9332-8_7, © Springer Science+Business Media New York 2013

radio. The resulting Babel of different waveforms made the Army very interested
in SDR. The largest single SDR development effort is spearheaded by the joint
tactical radio system (JTRS) joint program executive office (JPEO)[1] [93]. The
JTRS program started in 1997 and was still ongoing in 2011. Its stated goal is
development of a standard to facilitate reuse of waveform code between different
radio platforms.

The main standard developed by JTRS is the software communications architecture (SCA). SCA defines how waveform components are defined, created,
connected together, and the environment in which they operate. [94]. The standard
was developed with a very software-centric view—an SCA-compliant radio was
envisioned as a true SDR running on a GPP. Unfortunately, at the time SCA was
defined, GPPs were relatively slow and not up to the task of running a waveform.
SCA radios were perceived as very slow, inefficient, and inelegant. In the past
decade, GPPs have caught up with this vision and SCA implementations have
become more efficient. Just when JTRS started fielding actual devices, the JPEO
started an effort to update the SCA. A draft specification of SCA Next was released
in 2010.

7.1.1 SCA Background

A complete discussion of the SCA is beyond the scope of this book, and the reader
is encouraged to review [94]. The SCA defines:

- An operating system (RTOS) for the hardware that the radio runs on. The OS
  must be substantially POSIX compliant, implying availability of a reasonably
  powerful GPP (i.e., an 8-bit microcontroller is probably not sufficient).
- A middleware layer that takes care of connecting different parts of the radio
  together and handles data transfer between them. In the first version of SCA
  (2.2), the middleware layer had to be CORBA. CORBA is an industrial-grade
  software architecture construct that was developed to facilitate communications
  between software written in many different languages and executing across
  multiple computers. The SCA Next standard removes the requirement that the
  middleware be CORBA and leaves it up to the radio developer. Since all SCA
  radios at this time rely on CORBA, it will be assumed for subsequent discussion.
- A set of interface definition language (IDL) classes provide an abstract interface
  to different objects that make up a radio (e.g., IO ports, components, devices,
  etc.). These classes, together with software code that enables their use, are
  known as the core framework (CF).
- An XML ontology (Domain Profile) to describe all the components that make up
  a radio and how these components are to be interconnected and executed to
  implement the desired waveform.

---

[1] For conciseness, JTRS is used to refer both to the program and the office.

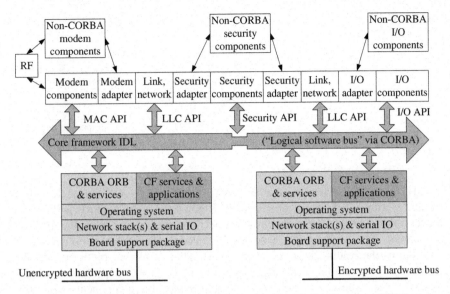

Fig. 7.1 Conceptual hierarchy of SCA components (lightly shaded blocks are COTS, darker shading is the SCA core framework, unshaded blocks are application-specific)

- APIs for many frequently used interfaces (e.g., audio, Ethernet, etc.)

A conceptual hierarchy of the different parts of SCA is shown in Fig. 7.1. The first takeaway is that SCA is a large and complex system. It is designed to address security requirements of military radios and has built-in support for red/black (secure/open) separation. The next few paragraphs will cover some of the major parts in more detail.

7.1.1.1 CORBA

A large and computationally intensive software program runs on many processors. Some of the processors are directly connected to each other (perhaps even on the same chip), while others are in a different chassis. Routines executing on different processors need to communicate with each other. This problem was first addressed in the context of supercomputers, and standard message passing techniques were developed (e.g., message passing interface (MPI)). MPI was developed for high performance, low-level programming and deals exclusively with data exchange. A higher level of abstraction is provided by the remote procedure call (RPC), which allows a programmer to call functions that may end up executing on a different processor. RPC requires the client (caller) and server (callee) to know about the existence of the function, its parameters, and the location of the server. The function and parameters are defined in a *stub*, which is analogous to a 'C' header

Fig. 7.2 Wrappers for a
client written in Python and a
server written in 'C'

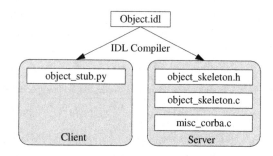

file. The operating system (or another layer that implements RPC) hides the details of the function call from the programmer. RPC on the client contacts RPC on the server (e.g., by opening a network connection), and sends over the function parameters. RPC on the server executes the function with the received parameters and returns the result. The next level of abstraction is offered by object-oriented distributed programming. Distributed object systems make it possible to design and implement reusable, modular, and easily deployable components with complexity hidden behind layers of abstraction.

CORBA is one implementation of a distributed object system [95]. The key to CORBA is an object request broker (ORB) that is responsible for data transfer between and execution of software objects residing on possibly different processors. ORB is similar to RPC but with many more features. To begin with, an object is described using interface description language (IDL)—specifying the inputs, outputs, and methods. Note that the object paradigm is slightly different than the function paradigm, but the inputs can be thought of as function parameters and outputs are values returned by the function. The IDL description is compiled by a tool provided by the developer of the ORB to create two wrappers: one for the client side, called a stub, and one for the server side, called a skeleton (Fig. 7.2). The client and server may use different languages and the IDL compiler can generate wrappers for each language (e.g., C, C++, Python, JAVA, etc.). The ORB itself may be written in any language. The ORB executing on each processor may come from a different vendor. The 'object_skeleton' contains auto-generated code to support data transfer, setting parameters, and calling of methods. CORBA supports powerful features that allow software objects to find each other at runtime by querying the domain name service. The ORB abstracts the communications medium. While most CORBA implementations rely on the standard TCP/IP network stack for low-level communications, this is not a requirement. ORB works just as well over PCI-express, or a custom hardware interface. A sophisticated ORB can detect when the two objects are executing on the same processor and use shared memory for communications. (Fig. 7.3)

Although IDL is the magic that lets users define consistent interfaces between blocks in an SCA radio, nobody actually writes IDL code. IDL descriptions are automatically generated by tools such as OSSIE that allow the user to think in terms of IO ports and data types that are meaningful to radio designers. IDL

Fig. 7.3 CORBA objects
communicating through the
ORB

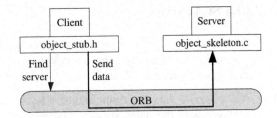

descriptions can also be generated from a UML drawing tool. UML is often associated with CORBA systems, since both are standardized by OMG. However, the two should not be confused as they are totally different things. Use of UML as a design language does not imply any particular architecture, or implementation language and it is not even restricted to software systems.

When SCA was first introduced, many developers were concerned with performance degradation caused by multiple layers of abstraction required for CORBA. These concerns are unfounded. The overhead due to data transfer at runtime is small as long as data is transferred in reasonable-sized chunks and each component performs a lot of signal processing on the data [96]. The overhead increases with the number of SCA components [97]. Not surprisingly, having many simple components is less efficient than having a few larger components. The overhead when the application if first started, on the other, hand, is quite large. A system CORBA requires a lot of setup time for the components to register with the ORB and then for the ORB to establish connections between components.

7.1.1.2 CORBA for Non-GPPs

CORBA is intended to be implemented on GPPs. However, many SDRs include non-GPP devices such as FPGAs and DSPs. These devices are not suitable for executing a full-featured ORB and therefore cannot fully participate in the ORB infrastructure. SCA addressed non-GPP hardware by defining an adaptor component. This component executes on the GPP and translates CORBA messages to modem hardware abstraction layer (MHAL) messages. The MHAL messages are simple packets with the packet format left entirely up to the developer.

The SCA Next standard specifies two new CORBA profiles:

- Lightweight Profile for SCA applications hosted on resource-constrained platforms such as DSPs
- Ultra-Lightweight Profile targeted at applications hosted on FPGAs

These profiles support a small subset of CORBA functions and data types. The ultra-lightweight profile can be implemented entirely in FPGA firmware (e.g., using VHDL).

7.1.1.3 SCA Services

Developers of SCA realized that most SDRs require a few common services from
the operating environment. These services were made a required part of the SCA
operating environment (OE):

- *Name Service* acts as a central registry for blocks to find each other. Each block
  registers with the name service. The application factory then uses XML configuration files to connect blocks.
- *Event Service* provides a mechanism for blocks to send and receive events. A
  block may register to receive events of a given type and the event service takes
  care of routing the events.
- *Log Service* provides a standard mechanism for blocks to generate timestamped
  record of events (e.g., application started at this time).
- *File Service* provides an abstract view of a filesystem. Many SDR developers
  are at first surprised that a file system is even a part of a radio. However, from a
  software perspective, a file system provides a standard way to exchange data and
  control (e.g., on Linux a file may very well be an interface to some hardware).

7.1.1.4 XML Ontology for SCA

SCA OE builds up SDR applications from constituent components at runtime.
A number of different XML files are used to describe the components and how to
interconnect them. Note that the XML files have nothing to do with the *functionality*[2]
of the components—just what the inputs and outputs are and what other modules are
connected. The hierarchical relationship between the various XML files is shown in
Fig. 7.4. In a nutshell, there are eight different XML files defined in SCA:

- Software component descriptor (SCD) describes a CORBA component's
  characteristics.
- Software package descriptor (SPD) describes a component implementation.
- Software assembly descriptor (SAD) describes an application's deployment
  characteristics.
- Property file (PRF) describes properties for a component.
- Device package descriptor (DPD) identifies a class of hardware device and its
  characteristics.
- Device configuration descriptor (DCD) describes configuration characteristics
  for a Device Manager.
- Domain manager configuration descriptor (DMD) describes configuration
  characteristics for a Domain Manager.
- Profile descriptor describes a type of file (SAD, SPD, DCD) along with the file
  name.

---

[2] See Sect. 7.3 for a description of an XML ontology meant to convey *functionality* rather than
connectivity.

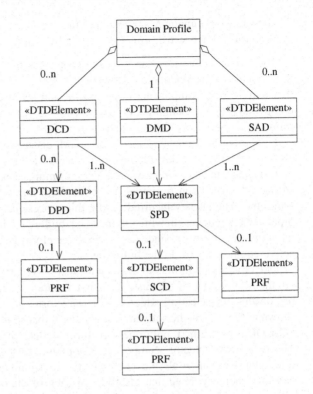

Fig. 7.4 Relationships between XML files for SCA [93]

The most interesting file is the 'software assembly descriptor' (SAD). The SAD file tells the OE about all the components that make up the SDR application and how they are connected. The 'device configuration descriptor' (DCD) file tells the OE about all the available devices and services (e.g., processors, FPGAs, etc.).

The user is unlikely to have to create any of these files by hand. Vendor-provided tools such as OSSIE (see Sect. 8.2) or SpectraCX allow the user to simply draw block diagrams representing the desired SDR application, which are then converted into SCA XML files. However, it is a good idea for SDR developers to have some notion of what is happening 'under the hood'. Readers interested in implementing an SCA SDR are encouraged to see Appendix C for a walkthrough of the XML files.

7.1.2 Controlling the Waveform in SCA

SCA provides two mechanisms to control an executing waveform:

1. Low-level interface to each component via the configuration properties accessed by the Configure method.

2. High-level (Domain Manager) interface to load, start, stop, and unload a
   waveform. Each waveform is described by a different SAD file.

The low-level interface can be used to control a few runtime parameters such as
the symbol rate or modulation. However, it is not suitable for large-scale changes
that require new components to be loaded and swapped for currently executing
components (e.g., changing FEC from Turbo to LDPC). Large-scale reconfiguration of an SCA waveform is addressed in [98].

Fast waveform changes (e.g., for ACM discussed in Sect. 3.1.1) were not
considered during the development of the SCA standard. Unloading one waveform
and loading another is relatively time-consuming, especially on a resource-constrained device. Even retrieving the code from non-volatile memory can take a
long time. The latency incurred during this process may be unacceptably high. If
input data continues flowing, it must be buffered while the reconfiguration is
taking place. However, since the original waveform is first unloaded, there are no
active components to do the buffering.

One obvious solution is to develop a super waveform that contains the code for
all the waveforms of interest. The first component in the super waveform is
responsible for selecting which of the waveforms receives data. This approach,
shown in Fig. 7.5, is the software equivalent of that shown in Fig. 2.1. As far as the
SCA OE is concerned, only one waveform, called 'Super-Waveform', is loaded.
The 'waveform selector' component is configured using the low-level configuration interface. Since all the components are simultaneously resident, they all
consume memory. Note that the code for each of the components is only loaded
once and can be shared among all the waveforms (i.e., if the same component is
used for phase tracking in all the waveforms, it is only loaded once). However,
each component does allocate its own runtime data. The inactive components use
negligible CPU resources. One disadvantage of this approach is that new waveforms cannot be easily added since the super waveform has to be rebuilt.

A more elegant solution, proposed in [98], requires extending SCA to support
dynamic changes to the connections between components. The CORBA middleware layer is still used to establish connectivity, but new functions are required
outside the Application Factory.

7.1.3 SCA APIs

SCA compliance is not sufficient for easy waveform portability. Consider an
application (waveform) that relies on a specific and unique audio interface that is
only available on one hardware platform. Even if the application is SCA compliant, it would be quite difficult to port to another platform. To address this
problem, the OE defines a set of standard interfaces to frequently used devices. A
layer of standardized APIs is placed between the application and hardware, as
shown in Fig. 7.6 [99, 100]. The APIs, just like the rest of SCA components, are

Fig. 7.5 'Super-waveform' approach to large-scale reconfiguration under SCA

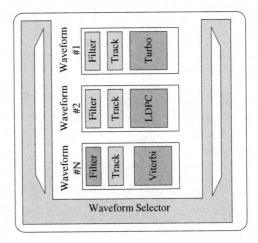

described using IDL. As shown in Fig. 7.6b, the hardware vendor must provide a Radio Device which translates between the JTRS API and the low-level, hardware-specific API.

A set of fundamental, abstract APIs define interfaces common to a wide range of more specific APIs. These primitive APIs are not intended to be used directly by the waveform developer[3]:

- The *Packet* interfaces provide methods to push two-way packets of common data types to the device or service. It also provides methods to configure and query the packet payload sizes.
- The *Device IO* interfaces provide methods to enable and signal Request-to-Send (RTS) and Clear-to-Send (CTS) messages.
- The *Device Message Control* interface provides control for communications between a receiver and transmitter. It allows the Device User to indicate when the transmission is active and also provides an interface for aborting the transmission.

The specific APIs are derived from the primitive APIs. The specific APIs defined as of 2011 are:

- Two audio interfaces are defined. The *AudioPort* device provides only the most basic audio output and is used to generate 'beeps'. The *AudioSampleStream* extends the functionality of *AudioPort* to provide the ability to consume and control audio samples to/from the audio hardware. This device would be used to interface to a speaker and microphone. Audio samples are exchanged as packets.

---

[3] A set of legacy APIs duplicate most of the functionality in *packet* and *device*. These APIs are maintained to avoid breaking legacy applications and should not to be used for new development: Device Packet, Device Packet Signals, Device Simple Packet, Device Simple Packet Signals, Device IO Signals, Device IO Control.

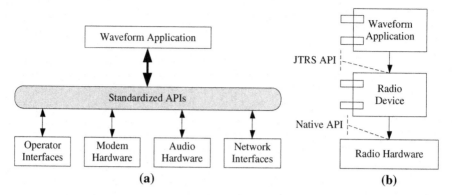

Fig. 7.6 Use of APIs in JTRS **a** conceptual view, **b** UML view

- Since many JTRS radios are meant for voice communications, a vocoder interface is one of the basic APIs. *Vocoder Service* API encapsulates vocoding capabilities that are common across all waveforms and applications. The Vocoder Service API supports loopback operations, provides for the transfer of encoded/decoded bit streams to and from the service user, and defines operations to select algorithms supplied by the vocoder. It should be noted that the use of the base Vocoder Service API and codec extensions require an implicit or explicit connection with the Audio Port device API. Implicit connections are defined by the platform implementation. Explicit connections are defined using the Vocoder Audio Stream API Extension. The vocoder service is one of the more complex APIs.
- The *Serial Port* device provides the ability to send flow-controlled byte packets to/from the device or service user and supports RTS/CTS handshaking. The Serial Port device API also provides a base configuration interface. A specific type of serial port is selected by using one of the extensions:

  - *Asynchronous* extension defines a UART interface
  - *Synchronous* extension defines a synchronous serial interface
  - *Synchronous High-Level Data Link Control* (HDLC) extension extends the functionality of the Serial Port to include synchronous serial capabilities using HDLC [101].

- The *Ethernet* device abstracts the interface to an Ethernet network. It closely mirrors the low-level packet interface[4] but adds Ethernet-specific addressing based on unique MACs.
- The *GPS* device provides an interface to get the user's position and/or time. The format of position and time information is set by deriving one of the two extensions:

---

[4] The Ethernet device is derived from the legacy Device Packet and DeviceIO rather than the new Packet API (see footnote 3).

- *Latitude/Longitude (LatLong)* extension for radian latitude and longitude coordinates.
- *Military Grid Reference System (MGRS)* extension for MGRS coordinates [102].

• *Timing* service maintains, manages, and distributes time within the JTR set. This includes Terminal Time and System Time management and distribution. The Timing service provides an interface for retrieving System Time and the quality indicator, i.e., Time Figure of Merit (TFOM), for Terminal and System Time. Terminal Time is the time returned from the OS and is monotonically increasing. Terminal Time is used for communicating time among the different terminal components (including distributed processor software and hardware components). The Timing Service synchronizes the Terminal Time between distributed components within the terminal. The Timing Service controls the local processor's POSIX clock. System Time is the terminal's estimate of Coordinated Universal Time (UTC) time. UTC time can be derived from various combinations of inputs (e.g., the GPS device, the chronometer device, or operator input) while utilizing the local timing pulse. Note that the timing service should be used in preference to getting the time from the GPS device since it is more general (e.g., timing may be available when GPS is not).

• *Frequency Reference* device provides the ability to read frequency reference data and to set the GPS Time Figure of Merit (TFOM) value.

• *Modem hardware abstraction layer* (MHAL) modem hardware abstraction layer abstracts the modem interfaces from the application software. The MHAL API supports communications between application components hosted on GPPs, DSPs, and/or FPGAs. This API is described in more detail in Sect. 7.1.1.2.

7.2 STRS

SDR is very attractive for use in satellites and deep space missions. The radios on space platforms have very long life cycles. As discussed in Sect. 3.6, changing the communications waveform can extend the life of a mission, increase the amount of returned data, or even save the mission from failure. NASA has been studying SDR for the past 10 years and has also realized the need for a standard. Hardware suitable for use in space lags behind commercial hardware by at least 10 years. Thus, flight radios are significantly more constrained in terms of computational resources than terrestrial equivalents. The SCA standard described in the previous section is too heavy for implementation on most NASA missions (especially deep space ones). NASA decided to develop a lightweight standard modeled loosely on the JTRS standard, and called it 'Space Telecommunications Radio System' or STRS [103].

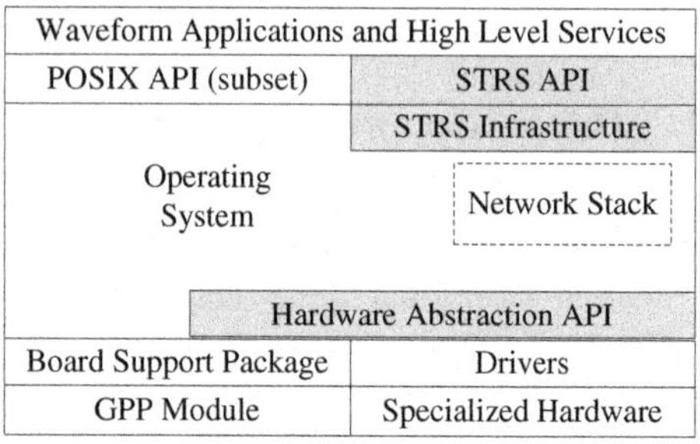

Fig. 7.7 Conceptual
hierarchy of STRS
components

Waveform Applications and High Level Services	
POSIX API (subset)	STRS API
	STRS Infrastructure
Operating System	Network Stack
	Hardware Abstraction API
Board Support Package	Drivers
GPP Module	Specialized Hardware

STRS has been in development since 2007 [92] and at least two flight experiments have been flown by 2011 with STRS-compliant radios. A conceptual
hierarchical view of STRS components is shown in Fig. 7.7. Comparing this figure
with Fig. 7.1 clearly demonstrates that STRS is significantly simpler than SCA.
Perhaps the largest difference is that no middleware layer (i.e., CORBA) is used
for STRS.[5] The bulk of the STRS standard deals with APIs. Unlike SCA, the
STRS standard also deals with hardware implementation of the radio. JTRS
developers were mostly concerned with application and waveform portability.
STRS developers also want to be able to swap hardware boards between radios or
add new ones.

STRS developers learned from the problems encountered by SCA developers
when integrating non-GPP components into the radio. STRS explicitly addresses
the 'signal processing module (SPM)' made up of ASICs, FPGAs, and DSPs that
do *not* execute software. In fact, the GPP is meant for control only and is not
expected to do any signal processing at all.

The set of APIs defined by STRS is much smaller than for JTRS. Most of the
required APIs deal with the infrastructure and would be considered a part of the
Application Factory in JTRS. The API classes are:

- [Application Control] includes commands to interact with the waveform.
  Configure, start, stop, test, etc. This API is defined twice—one for the waveform
  level and one for the infrastructure level. The user only interacts with the
  infrastructure, which in turn calls the waveform-level commands.
- [Application Setup] includes commands to prepare the application for execution. Instantiate, remove, upload, etc.
- [Device Control] includes commands to interact with the hardware abstraction
  layer. The HAL is not yet defined by STRS. Open, close, start, stop, etc.
- [Memory API] includes commands for generic memory allocation. Memory is
  often a very scarce resource on flight radios and must be managed carefully.
  Further, NASA guidelines discourage dynamic memory allocation using the

---

[5] STRS does not require a POSIX-compliant operating system or CORBA, allowing
implementation of small- and low-power STRS radios.

standard 'malloc' function. Only three functions are defined in this API: copy, reserve, and release.
- [Messaging API] includes commands for a simple message passing architecture. Queues are used to send messages between threads in an application or between applications. Both basic and subscriber-based message passing is supported.
- [Time Control] includes commands to get and set time.

The STRS standard is still a work in progress. STRS-compliant radios are likely to facilitate easier waveform and application portability, but definitely do not guarantee it.

7.3 Physical Layer Description

The SCA and STRS standards deal primarily with waveform portability and provide a waveform development environment. The actual waveform (physical layer) is considered a black box[6] that provides a means for sending or receiving bits. No open standard that addresses the physical layer is currently available. Consider the simple task of specifying the signal carrier frequency—some refer to this parameter as f_c, others as *carrier*, and others as *freq*. Multiple researchers have identified the need for such a standard [104,105,106], but their proposals did not come to fruition and have not been accepted by the community. Another standard, SDR PHYsical layer description (SDRPHY), was proposed by the author of [107] to fill the gap by providing a lightweight description language that covers the *samples-to-bits* part of the SDR. A crude analogy is: "SCA and STRS provide the building codes and some of the materials. SDRPHY aims to provide the blueprints. Both are required to build a house". As of 2011, SDRPHY has not been accepted by the SDR community or used beyond a few experimental demonstrations. A website, http://sdrphy.org, has been established to allow open access to and collaborative development of the standard.

A common language for describing waveforms for wireless communication can:

- Facilitate reuse of software developed for different hardware platforms.
- Reduce the cost and time to move between different vendors' solutions.
- Stimulate research in software-defined and cognitive radios by simplifying collaboration between different teams.
- Reduce procurement costs by encouraging competition between vendors
- Help students and researchers see the commonalities between different waveforms rather than focusing on the unique details.

---

[6] The Application Factory in SCA does not know or care about the intent of the component connections defined in the SAD file.

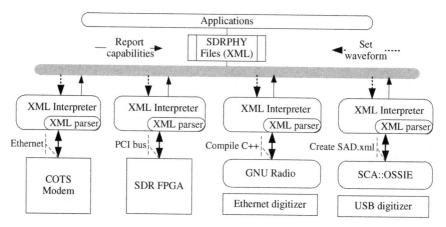

Fig. 7.8 Conceptual use of the XML standard

Many of these goals are similar to the goals of SCA. However, SDRPHY is not an alternative to SCA, but is in fact complementary.

A software program, called an *interpreter,* provides an interface between SDRPHY and a specific SDR implementation. The interpreter converts the description into a set of configuration commands, or creates a flowgraph (e.g., an SCA SAD file), or even generates code to implement a radio satisfying the description. The interpreter may reside within the SDR hardware, or operate entirely offline. The interpreter should be able to:

- Report hardware capabilities.
- Report current configuration.
- Configure or reconfigure a set of options within the hardware capabilities.
- Report current operating environment.
- Provide a hardware-agnostic method to transmit and receive data.

The configuration, capabilities, and status are described using XML. A conceptual block diagram showing four different radios is given in Fig. 7.8. The main concept illustrated in the figure is a hardware-agnostic layer that allows multiple applications to work with different SDR hardware. An application, which could be anything from a graphical user interface to a cognitive engine, generates XML configuration files that are then passed to a hardware-specific interpreter.

These concepts are easily extensible to support cognitive radio, which is a superset of SDR. The primary use of the standard is to exchange configuration information between the application and the radio. The two are assumed to be co-located or have some pre-established reliable means of communication. However, the standard can also be applied (with minor changes) to allow over-the-air exchange of capability information for cognitive radio negotiation. The standard does not address how an initial link between two cognitive radios is established.

The XML format used to configure an SDR can also be used to describe users in a radio channel. For example, a cognitive radio detects RF emitters in the area and reports them as a set of SDR configurations.

7.3.1 Use Cases

Consider a ground station responsible for control and monitoring of multiple satellites. Each satellite may use a different waveform. Currently a set of configuration files in a proprietary format is used to setup the ground transceivers for each satellite. A new satellite transponder is supported by developing a new set of configuration files. Since the format is proprietary, it falls on the ground station operators to develop and test the configuration files. The satellite developer, who is most familiar with the transponder, is unable to help. An agreed-upon standard for waveform description would facilitate communication between them.

Consider a cognitive radio engine that can optimally select a waveform and carrier frequency under different channel conditions. Developing such an engine is a major undertaking. However, if the software is designed for a particular SDR hardware platform, another major effort is required to port it to other platforms. A waveform description standard allows application portability. Further, two radios may exchange their 'waveform capability' descriptions and use an automated method to determine the best mutually supported waveform. Currently, there is no standard way to exchange this information.

Consider a monitoring receiver that processes many different waveforms. Adding a new waveform currently requires knowledge of the receiver user interface. Again, a standardized waveform description language is needed.

7.3.2 Development Approach

The goal of this standard is to provide a lightweight description language that covers the *samples-to-bits* part of the SDR. It is not meant to replace JTRS/SCA or STRS but to augment them. In fact, configuration of an SCA-based SDR using SDRPHY is described later in this chapter. The transmitter functionality starts with data bits and ends at the antenna, while the receiver starts at the antenna and ends with data bits. In other words, only the physical layer is considered.

The standard should ideally satisfy the following goals

- *Completeness* Most practical waveforms should be describable using the standard. The goal is for 99 % of all waveforms of interest to the user community to be supported. The remaining 1 % of exotic waveforms may be overlooked because it would make the standard too complex.

- *Lightweight* The interpreters for the standard are meant to be implemented in a wide range of devices—from satellites to handheld units. Therefore, low processing and memory requirements are desirable.
- *Consistency* The standard should be self-consistent. Similar functionality should not require different descriptions. A high-level, abstract view of the waveform is adapted whenever possible. Conceptual commonalities should be exploited.
- *Compactness* The same keywords should be used for describing the configuration and capabilities, whenever possible. In particular, no new keywords should be created if existing ones can be repurposed.
- *Accessibility* There should be no barriers to entry for users wishing to participate in the creation or utilization of this standard.

The consistency and compactness goals are currently being debated. There is a tradeoff between using the most abstract view of a parameter and ease of implementation and specification. Consider a Reed-Solomon (RS) error correction code—it can be shown that RS is a special case of a BCH code [108]. Therefore, it is sufficient to define a tag for BCH. On the other hand, many users of the standard may not be aware of the equivalency between RS and BCH, placing an undue burden on them. One approach is to provide Reed-Solomon as a library entity, based on BCH. Similar tradeoffs come up in many other cases.

The goal of this standard is to enable application developers to work with a wide range of *practical* flexible radios. The idea of a practical radio limits the scope of the standard: it does not attempt to describe an arbitrary radio. Some combinations of features may be physically possible, but are either impractical or highly unusual. For example:

A waveform may use either convolutional (Viterbi) or LDPC forward error correction. It is possible to concatenate Viterbi and LDPC, but such a combination makes little sense from the communications architecture perspective. Likewise, it is possible to apply forward error correction to the *chips* in a spread-spectrum system instead of applying forward error correction (FEC) to the *symbols*. However, no practical radio uses this technique.

The standard can therefore rely on a large set of *implied constraints*, based on a *canonical* communications system. These constraints are explicitly stated whenever multiple interpretations of the XML are possible.

The implied constraints and the concept of a canonical system are an integral part of the interpreter. It shifts some of the complexity of specifying a waveform from the application designer to the interpreter. For example, the application designer should be able to construct XML code to configure a radio by reading a published standard. The complexity of such an interpreter can be a major impediment to adoption of the standard. The author implemented three such (admittedly limited) interpreters to demonstrate feasibility, as discussed in Sect. 7.3.5. The same configuration file is passed to all radios and the radios can exchange data with each other.

There are at least two approaches to describe a waveform, but neither is fully satisfactory:

1. Define constituent blocks and interconnections between them (apparently adapted by [105] and the SAD files for SCA). This approach comes close to specifying the implementation rather than the intent. Once the block-based paradigm is allowed, there is no logical reason why it cannot be extended to specify every adder and multiplier in the system.
2. Specify values for all relevant system parameters (this is the approach used by most published waveform standards). This approach is problematic due to possible ambiguities. For example if an FEC encoder and an interleaver are both described, it is not clear in which order the operations should be applied to the data.

The SDRPHY standard is mostly based on the second approach. Whenever possible, the goal is to describe the *intent* rather than the *implementation*. Most of the interconnect is implied by the canonical communications system. However, connectivity can be specified when absolutely necessary. Any constituent block described in XML can be uniquely identified with an *ID* attribute. Connectivity is specified by adding a <input ID=''unique_id''> node to the block following the uniquely identified block.

Unique aspects of some systems cannot be adequately described using the XML tags defined in this standard. For example, a spreading sequence for a DSSS system can be based on a proprietary cryptographic generator (e.g., GPS military codes). Describing such a generator is well beyond the scope of this standard. A *foreign* attribute is therefore defined to allow the user to supply a non-standard component. The *foreign* attribute is very problematic since it must be supported across different platforms—from software to hardware. The component interface must therefore be very well-defined and should be based on existing industry standards (e.g., CORBA for software and Open Core Protocol (OCP) for hardware).

7.3.3 A Configuration Fragment

A comprehensive review of SDRPHY is beyond the scope of this book. Instead, one example is provided below and the reader is encouraged to explore the rest of the standard at http://sdrphy.org.

Most practical communications systems impose a packet and/or frame structure on the transmitted symbols. Let us consider an appropriate XML description for the physical layer framing. The framing may be taken care of at the bit level (e.g., older SATCOM signals), or may have to be managed at the symbol level (e.g., DVB-S2). Some standards use different modulation and/or coding for different parts of the frame.[7] The frame structure for these standards must be described

---

[7] This is also the case for OFDM waveforms, where the frame synchronization sequence is not necessarily an OFDM symbol.

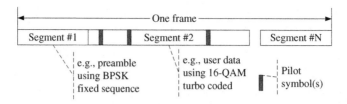

Fig. 7.9 Canonical wireless data frame structure

before a transmitter or a receiver can be created. A frame structure that can be handled in the receiver at the data bit level (after hard decisions) need not be described in SDRPHY.

A frame can be described as a sequence of fixed-duration segments (a segment is the largest set of samples that can be processed at once). Each segment may be described by a different waveform (e.g., a preamble may use a robust modulation such as BPSK, while the bulk data uses an advanced modulation). Most frames consist of just two segments—preamble and bulk data. The preamble is typically (but not always) a combination of a fixed pattern and a short user-defined sequence. In particular, an FEC block must be contained within a single segment. (Fig. 7.9)

A frame is defined in XML as a sequence of segments. As with any XML for this standard, the waveforms may be defined inside the segment definition, or reference previously defined waveforms. Three variables must be defined for each segment (see Sect. 7.3.7 for an example of their use):

- <duration>. Set the duration of the frame segment (in symbols) to be transmitted using the selected waveform.
- <data_source>. The data source selects one of: a named data source, a file, or a fixed data sequence.

  – A named data source provides a standard way for the application to supply data to the hardware.

  - <waveform>. Define all the parameters of the waveform to be used for the segment, such as: data rate, modulation, coding, etc.

Many waveforms insert known (pilot) symbols throughout the frame to help with acquisition and tracking. The pilots are inserted after FEC and can be thought to exist at the 'frame' rather than at the 'segment' level. The pilots are described by their position in the frame and the symbol values at each position. SDRPHY reuses the concept of a segment to describe each block of pilots (frequently just one symbol). The <positions> tag defines offsets for each set of pilot symbols relative to the first symbol in the first segment. If all the pilots are the same, only one tag is used. However, if the pilots at different positions are not all the same, a separate tag is required for each position. The frame hierarchy is conceptually shown in Fig. 7.10.

Fig. 7.10 SDRPHY
hierarchy to describe a typical
waveform frame

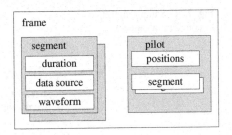

7.3.4 Configuration and Reporting XML

XML has a number of advantages, the main ones being widespread acceptance and human readability. XML does come at a price since it is rather verbose,[8] and therefore requires more resources to generate, store, and parse than a custom binary format. However, this overhead is negligible, given the complexity of today's SDRs.

An XML description consists of a hierarchical tree of nodes. Each node represents a conceptual part of a waveform. The hierarchy is not always obvious. In general, node B is a child of node A if it satisfies the following question: Is B a property of A? For example: Does a radio consist of multiple transmitters and receivers, or does each transmitter consist of multiple channels?

The top-level node is <radio>. The XML standard defines a companion constraint language—an XML Schema Definition (XSD). The SDRPHY standard is itself an instance of an XSD file. Leveraging XML and XSD allows the designers to take advantage of a rich set of development tools. In fact, an XSD file can be used to automatically generate a user interface suitable for defining waveforms. SDR capabilities can also be reported as an XSD file by providing a set of constraints on top of the SDRPHY standard. For example, the standard specifies that <modulation> can be either <linear> or <OFDM>. A capabilities report may restrict <modulation> to be only <linear>. Existing XML parsers can be used to verify that a given configuration file is valid for a given capabilities file.

A set of standard libraries (e.g., defining 'BPSK', 'rate 1/2 convolutional encoder with K = 7') are provided. These libraries can be used to create compact XML descriptions for common waveforms. A radio may be queried to list all the libraries that it has already stored. Some radios may be able to save the uploaded libraries so that they do not need to be resent with every XML file.

---

[8] XML files can be compressed using the Fast Infoset (FI) binary encoding standard [255]. The FI files may be further compressed using zlib. See [256] for a comparison of different XML compression strategies.

Fig. 7.11 Sample flowchart
to configure a modem

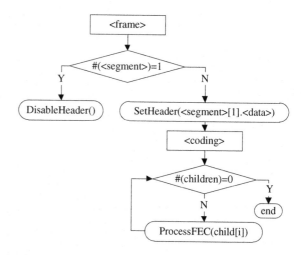

7.3.5 Interpreters for Hardware-Centric Radios

Two proof-of-concept interpreters were developed to demonstrate the feasibility of
using SDRPHY to configure a radio. The Comtech [109] modem uses SNMP to
configure the hardware. The interpreter is a stand-alone program developed in C++
that takes in an XML configuration file and outputs a stream of SNMP commands
to the modem. The capabilities file for the modem is created manually since the
number of supported features is not very large.

The FPGA-based SDR is configured through a set of proprietary API calls. A
large library of functions is used to configure everything from data rate to forward
error correction to the frame structure. The number of options is larger than for the
modem, but still quite manageable. The capabilities file is once again created
manually.

For both the FPGA and the COTS modem, the interpreter follows a hard-coded
flowchart to determine the appropriate SNMP commands or API calls. Very little
error checking has to be done by the interpreter since an XML validator verifies
that the configuration is within the radio capabilities. A fragment of a flowchart is
shown in Fig. 7.11. At every node in the XML tree, the interpreter checks if it has
to issue an API call, or delve deeper before a call can be made.

7.3.6 Interpreters for Software-Centric Radios

An interpreter for an all-software radio is significantly more complex than for a
COTS modem. For example, GNURadio framework (see Sect. 8.1) allows creation
of an essentially infinite number of different radio architectures. An advanced
interpreter may even create new functional blocks in C++ to satisfy the

configuration file. However, the only requirement placed on any interpreter is that it be able to configure a radio given an XML configuration file that satisfies the capabilities report. The capabilities report is generated by the interpreter itself.[9] Interpreters for two popular SDR frameworks, GNURadio and OSSIE, were created to demonstrate feasibility.

In GNURadio, each signal processing block (e.g., equalizer, phase locked loop) is defined as a C++ class. The *connect* method of each class is called to create a full receiver flowgraph. An SDR was designed using GNURadio framework by defining several configurable blocks and a flexible connection between blocks. In order to be SDRPHY compliant, an interpreter was written in C++ to parse an XML file and then directly set the configuration variables for the SDR. The interpreter generates a unique build script that selects the subset of features set in the configuration XML by defining preprocessor constants. For example, it sets the modulation and selects one of two FEC codecs. The front-end digitizer is configured based on the <carrier_frequency> and the <symbol_rate> tags.

OSSIE is an SDR framework for implementing key elements of the SCA specification (see Sects. 7.1 and 8.2). OSSIE is used to create *waveforms* consisting of multiple *components* that complete the radio. A component is a processing block that performs a specific function with the provided input. A component can have properties set by the user. For instance an amplifier block has a property specifying the amount of amplification. SCA defines two XML files (.sad, .prt) to determine how the components are connected and the property values for each component.

An SDRPHY interpreter was developed in Python to generate these two XML files. The interpreter begins by parsing the XML configuration file and keeping track of the tags it parses. Once it recognizes a tag that requires a component that component is added to the .sad file. It continues to interpret the children and attributes of that tag to set the appropriate properties for that component. The process will continue until all tags have been parsed. At this point no connections between the components have been made yet. Connectivity between the components is implied by the canonical radio flow graph. If a component was not instantiated during the parsing of the configuration file, it will be left out of the chain and no connections will be made to it.

The functionality of these interpreters was demonstrated by having each radio relay a stream of packets over an RF connection to the other radios. The same SDRPHY configuration files were used to configure the radios for different modulations, data rates, and FEC.

---

[9] Incidentally, this approach limits the 'vaporware' claims of some SDR vendors that overstate the capabilities of the radio. While it is true that firmware can be written to implement more waveforms, the SDR as-sold only supports a finite set.

7.3.7 Example

This section provides a relatively complete example demonstrating an SDRHY description of a DVB-S2 transmitter. Comments in the table apply to the XML code above them. (Table 7.1)

7.4 Data Formats

As long as the SDR is implemented by a single developer and is integrated on a single board or chip, the format of the data exchanged between blocks is not important. In a hardware-centric architecture, digital samples can be exchanged on parallel buses. In a software-centric architecture, digital samples can be stored as arrays of appropriate data types. Data format becomes important when the processing is distributed between different entities and the entities are not co-located. Many radios implement mixed-signal processing and DSP on different chips. Consider a system shown in Fig. 7.12, where the analog signal is digitized in block A and then processed in block B. At a minimum a standard is desirable for transfer of data between these subsystems.

What if the data has some *metadata* attached to it? A typical example is a timestamp associated with a sample of data. Enforcing the data type between the sender and the receiver is not sufficient if data types are not uniform (e.g., samples are floating point and timestamps are integers). Of course, ad hoc solutions can be developed, where a separate connection is used to send metadata.

At least five standards to describe data exchange between SDR blocks are currently available (Table 7.2). Most of the time the data contains samples digitized at an intermediate frequency (IF, see Sect. 9.1.3). This proliferation is addressed and justified in [110]. Each standard targets a narrow segment of the market and implements only the minimal feature set required for that segment. High volume products such as RF front ends for handheld devices are extremely price sensitive and any extraneous features would make a product less competitive. However, as transistors keep getting cheaper, the incremental cost of implementing a full-featured standard decreases. Multiple standards will most likely consolidate or converge to just one or two.

7.4.1 VITA Radio Transport (VITA 49, VRT)

Modern military distributed systems rely on standard network infrastructure for all communication. Transfer of digital samples over a network can be standardized by requiring a middleware layer such as CORBA (see Sect. 7.1.1.1). A middleware layer can be responsible for ensuring that the data type sent by the source is the

Sample configuration file

```
<radio>
  <!-- Specifying a waveform for large frame, rate 1/2, 16-APSK -->
  <tx>
    <frame>
      <symbol_rate>1e6</symbol_rate>
```

Defining a transmitter frame structure. Symbol rate is 1 MSps

```
      <!-- Start of frame is 16 symbols-->
      <segment>
        <duration>16</duration>
        <data_source>
          <stream_sequence>
             1 0 1 0 0 1 1 0 1 0 0 0...
          </stream_sequence>
        </data_source>
```

The first segment is a 16-symbol preamble. The preamble sequence is
'101001101000...'

```
        <waveform>
          <modulation>
            <linear>
              <pulse_shaping>
                <raised_cosine>
                  <rolloff>0.3</rolloff>
                  <root>true</root>
                </raised_cosine>
              </pulse_shaping>
              <constellation>
                <points xref="standard_waveforms.xml?BPSK"/>
                <rotate>3.14</rotate>
              </constellation>
            </linear>
          </modulation>

        </waveform>
      </segment>
```

Root-raised cosine shaping filter with a rolloff of 0.3 is applied. The preamble is
sent using a π-BPSK modulation, where a regular BPSK constellation is rotated
by π every symbol. An optional <rotate> tag modifies a standard BPSK
constellation.   The definition for the BPSK constellation is referenced from a
standard library

```
      <!-- PLSCODE is 64 symbols -->
      <segment>
        <waveform>
          <modulation> ...  </modulation>
          <coding>
            <Reed-Muller>
              <order>1</order>
              <code>64,7,32</code>
            </Reed-Muller>
            <scrambler>
              <arbitrary>0 4 12 13 19 ... </arbitrary>
            </scrambler>
```

```
        </coding>
      </waveform>
      <duration>64</duration>
      <data_source>pls_code</data_source>
```

The second segment is a highly coded bit sequence that defines modulation and coding for the rest of the frame (see section 3.1.1 for details). The modulation is the same as the preamble and is shown as '...' for conciseness. Unlike the preamble, the data is coded with a combination of Reed-Muller and an interleaver. The interleaver is non-standard and is described by a complete lookup table. The <data_source> is a named stream, allowing the application to supply data for each frame.

```
<!-- DATA is 64800 bits, 16200 symbols-->
    <segment>
      <waveform>
        <modulation>
          <linear>
            <pulse_shaping> ... </pulse_shaping>
            <constellation>
              <points>
                <re>0.30</re><im>-1.14</im>
                <re>0.31</re><im>-0.31</im>
                  ...
                <re>-1.14</re><im>0.30</im>
              </points>
            </constellation>
          </linear>
        </modulation>
        <coding>
          <LDPC>...</LDPC>
          <BCH>
            <msg_len>32208</msg_len>
            <prim_poly>65581</prim_poly>
            ...
          </BCH>
          <interleaver>
            <block>
              <rows>16200</rows>
              <columns>4</columns>
            </block>
          </interleaver>
        </coding>
      </waveform>
      <duration>16200</duration>
      <data_source>data</data_source>
    </segment>
  </frame>
 </tx>
</radio>
```

The last segment contains the payload data. The modulation is defined by a set of 16 points in a complex plane. These points describe a 12/4-APSK constellation. Data is encoded with an LDPC block code, followed by a BCH block code. A subset of BCH polynomials is shown for conciseness. A standard block interleaver is used.

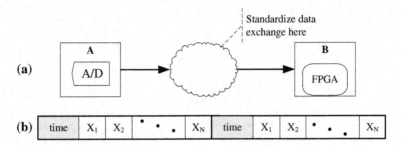

Fig. 7.12 Transmitting digitized data over a network **a** distributed computing demonstrates the need for standard data format, **b** representative standard with timestamped data

Table 7.2 A summary of digital IF standards

	Markets	Supported standards	PHY	Cost
VITA 49	Military/space	Agnostic	N	$100
digRF	High-volume chips	Agnostic	Y	$8000
SDDS	Military/intelligence	Agnostic	N	?
OBSAI	Cellular basestations	GSM/EDGE, WCDMA, LTE, WiMAX	Y	Free
CPRI	Cellular basestations	WCDMA, WiMAX, LTE	Y	Free

same as that expected by the sink (i.e., a floating-point value is not sent to a block expecting an integer). However, middleware is often not desirable (because of software complexity or overhead) or even feasible.

A platform- and middleware-independent standard for packetizing digital samples and attaching metadata to these samples is desirable. The VITA Radio Transport (VRT or VITA-49.0) protocol is an emerging standard for SDR applications targeted at military users [111, 112, 113]. It was developed to provide interoperability between a diversity of SDR components by defining a transport protocol to convey digitized signal data and receiver settings. Unlike the SCA standard, the VITA 49 is not particularly complicated or overarching. It only defines how data is to be packed into packets and states nothing about how it is to be processed. It is not the only standard that addresses this issue, but is apparently winning the battle of the standards, as evidenced by the availability of hardware [114].

The VRT specification was developed for receivers only, and does not explicitly support transmitters. This may appear as a major shortcoming but in practice is not. Parameters required to describe a receiver are usually a superset of those required for a transmitter (e.g., power level applies to both received power level and the transmit power).

VRT is meant to replace a set of analog IF cables and complicated analog switch matrices with a standard and low-cost digital network running IP. The vision [111] is encompassed in Fig. 6.34.

Table 7.3 VRT packet types

Contents	Standard formats	Custom formats
Data	*IF Data Packet* Conveys a digitized IF signal • Real/complex data • Fixed/floating-point formats • Flexible packing schemes	*Extension Data Packet* Conveys any signal or data • Any type of data • Custom packet format
Context	*IF Context Packet* Conveys context for IF Data • Frequency • Power • Timing • Geolocation • etc.	*Extension Context Packet* Conveys additional context • Any kind of Context • Custom packet format

Two types of packets are defined (Table 7.3):

1. Data packets carry digitized samples and timestamps associated with the samples. These packets are sent as data becomes available.
2. Context packets carry control and monitoring information. These packets are sent when the information changes (e.g., request to tune to a different frequency) or at a specified rate. Context packets are sent less frequently than data packets and require little network bandwidth. The context information is timestamped so that changes can be precisely related to the associated data.

VRT also supports the transfer of other signal data in a customizable transport format. This VRT protocol 'extension' capability supports any type of data that needs to be conveyed to handle a wide range of applications. For example, a VRT data packet can be used to send soft decisions from a demodulator to an FEC decoder. VRT defines the term 'information stream' as a set of related packet streams, typically at least one data stream and one context stream.

In order to associate the emitted VRT packets with VRT packet streams, each emitted packet contains a stream identifier. The stream identifier is a unique number assigned to a packet stream. The number is identical in all the packets of a given packet stream. Stream identifiers are also used to associate together all of the packet streams relating to an information stream. The context packets can be associated with data packets in five different ways (data/context, source/context, vector/context, asynchronous-channel/context, system/context), but only the data/context association will be discussed in this section.

7.4.1.1 Data Packet Format

The VRT packet format is extremely flexible. In fact, only a 32-bit header is required. The header specifies any additional header fields. In particular, the header specifies packet type (data, context, etc.), whether a stream identifier is

31	30	29	28	27	26	25	24	23	22	21	20	19	18	17	16	15	14	13	12	11	10	9	8	7	6	5	4	3	2	1	0
0	0	0	1	0	0	0	X	X	1	0	0	1	0	0	0	1	0	0	0	0	0	0	0	0	0	0	0	0	1	1	0

| Packet type | | | | | | | | TSI | TSF | | Packet #1 | | | | | Packet length = 6 words | | | | | | | | | | | | | | | |

| 0 | 0 | 0 | 0 | 0 | 0 | 0 | 0 | 0 | 0 | 0 | 0 | 0 | 0 | 0 | 0 | 0 | 0 | 0 | 0 | 0 | 0 | 0 | 0 | 1 | 0 | 1 | 0 | 1 | 0 | 1 | 1 |

Stream ID = 0x55

| 1 | 0 | 1 | 0 | 1 | 0 | 1 | 1 | 0 | 1 | 0 | 1 | 0 | 1 | 1 | 0 | 1 | 0 | 1 | 0 | 1 | 1 | 0 | 1 | 0 | 1 | 0 | 1 | 1 | 0 | 1 | 0 |

Time stamp in GPS format = 0x55555555 seconds

| 0 | 0 | 0 | 0 | 0 | 0 | 0 | 0 | 0 | 0 | 0 | 0 | 0 | 0 | 0 | 0 | 0 | 0 | 0 | 0 | 0 | 0 | 0 | 0 | 0 | 0 | 0 | 0 | 0 | 0 | 0 | 0 |
| 0 | 0 | 0 | 0 | 0 | 0 | 0 | 0 | 0 | 0 | 0 | 0 | 0 | 0 | 0 | 0 | 0 | 0 | 0 | 0 | 0 | 0 | 0 | 0 | 0 | 0 | 0 | 0 | 0 | 1 | 0 | 0 |

Time stamp in picoseconds is 4 ps after the GPS time in the previous field

| 0 | 0 | 0 | 0 | 0 | 0 | 0 | 1 | 0 | 0 | 0 | 0 | 0 | 0 | 1 | 0 | 0 | 0 | 0 | 0 | 0 | 0 | 1 | 1 | 0 | 0 | 0 | 0 | 0 | 1 | 0 | 0 |
| First sample = 1 | | | | | | | | Second sample = 2 | | | | | | | | Third sample = 3 | | | | | | | | Fourth sample = 4 | | | | | | | |

Fig. 7.13 Typical VRT data packet (TSI—integer timestamp, TSF—fractional timestamp)

available, and what type of timestamp is available. The packet length can be up to 256 kB in 4-byte increments. A typical VRT data packet format is shown Fig. 7.13.

Samples can be packed into 32-bit words either contiguously (with samples potentially split across two words), or using padding to ensure an integer number of samples per word. Each sample consists of the actual digital sample and two optional qualifiers that describe an event[10,11] that occurred at that sample.

The digitized samples can be either real valued or complex. Complex-valued samples can be expressed as either $a + ib$ or $re^{i\theta}$. Each component can be expressed as either fixed- or floating-point number and either signed or unsigned with 1—64-bit precision. Note that none of this formatting is explicitly described in the packet header. Two mechanisms are provided to specify the sample formatting:

1. An optional field called a class identifier selects a format from externally provided documentation (i.e., a vendor has to tell the user that packet class X means samples are 8-bit signed complex values).
2. A context packet (see next section) may contain a data packet payload format field ([111], 7.1.5.18) that describes the formatting.

It is interesting to compare the *almost* arbitrary format of a VRT packet to the strictly specified packets in the next subsections. The (overly) flexible nature of the VRT packets is a major impediment to the adoption of this standard outside the 'cost is no object' world of military users. Indeed, two systems that are both

---

[10] Events can also be indicated in an optional trailing word ([111], 6.1.7) or in a context packet.

[11] Sample format can have a significant effect on the performance of the GPP-based SDR. GPPs are designed to work with a few standard data types (e.g., 8-bit chars, 16-bit shorts, 32-bit integers, 32-bit floats, and 64-bit doubles). Further, high performance is achieved only when working on blocks of data rather than single samples. Interleaving flags and data samples require the GPP to 'unpack' the data before it can be operated on.

VITA 49 compliant may be unable to exchange data if they do not support every possible variation of the packet format.

7.4.1.2 Context Packet Format

Context packets are used to convey metadata that describes samples in the data packets. Context packets are not needed in the simplest systems. When context packets are used, the stream identifier is used to link data packets to the corresponding context packets. VRT defines 24 types of metadata (e.g., signal bandwidth and carrier frequency) that can be sent in a standard context packet.

One of these 24 types describes the data packet payload format and is critical for interpreting the associated data packets. This field describes formatting (e.g., unsigned 8-bit fixed point) and packing (e.g. continuous) of the sample data. If class identifier is not used to interpret the data format, at least one context packet must be received before any data packets can be processed.

Context packets can be transmitted without any associated data packets. For example, an RF downconverter can send out a context packet indicating its carrier frequency. The output of the downconverter is analog. A digitizer (ADC) then converts the analog signal to digital samples and generates VRT data packets. The ADC may also generate its own context packets, indicating the sample rate. The user can form a complete picture of the system by considering context packets from both the downconverter and the digitizer. As discussed previously, the combination of the two context packet streams and the data packet stream is called an information stream.

7.4.1.3 Packet Loss Mitigation

VITA 49 does not specify the physical or MAC layers. Standard Ethernet network infrastructure is envisioned for most implementations. The packets may be lost while traversing multiple switches and routers. The effect of a lost packet on the overall system performance depends on the details of the system. Infrequent packet loss may be booked against the target end-to-end BER and ignored, or it can be catastrophic if an entire user transmission is lost. MAC layer retransmission can be employed if the additional latency is tolerable (e.g., using TCP/IP) and the hardware is fast enough to support TCP/IP at the packet rate. To the best of the author's knowledge, all VITA 49 systems use the unreliable UDP protocol rather than TCP. Packet-level forward error correction can be used to compensate for lost packets at the cost of slightly higher throughput requirements [115]. Timestamps embedded in the packets can be used to detect dropped and out-of-order packets.

Fig. 7.14 digRF physical layer

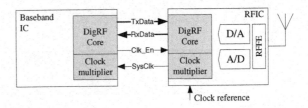

7.4.2 Digital RF (digRF)

The digRF standard is developed by MIPI alliance [116] to address the requirements of high-volume commercial products for the broadband wireless market. This specification defines the interface between one or more DSPs and RF front ends in a single terminal. The interface is intended to be efficient, flexible, and extensible, accommodating many variations in the overlying system design, while providing interoperability at the interface level between compliant ICs. digRF specifies everything from the names given to IO pins on the chips to the voltage levels used to the packet formats. This level of detail is needed to ensure compatibility between different vendors. The VITA49 standard described in the previous section addresses only the packet format, and is agnostic to the physical layer. Standard specification is made available only to MIPI members at a cost of about $10 k. Therefore, it is not appropriate for R&D institutions. The limited information available in the public domain and not under NDA comes from [117]. Six IO pins are defined for bidirectional data transfer between the chips, as shown in Fig. 7.14. Tx and Rx data are transferred serially using high-speed (up to 1.5 Gbps) differential signaling. The high line rate is derived from a lower frequency reference clock (typically 26 MHz), using a PLL inside both RF and DSP chips. The maximum line rate supports signal bandwidth up to about 75 MHz.[12]

Data are transferred in packets of up to 64 kB in size. Each packet can carry either I/Q samples or control information. This approach is similar to that used by VITA49.

7.4.3 SDDS

SDDS is an old format for the exchange of digitized sampled signals used widely in the military community. It is now being replaced by VITA 49 and is only described here for completeness.

SDDS packets are always encapsulated inside standard UDP packets and sent over the Ethernet networks [118]. The packet format is shown in Fig. 7.15. SDDS offers a few unique features:

---

[12] $(1.5 \times 10^9$ bps$)/(8$ b/sample$)/(2$ samples/Hz$)$ * packet overhead ~ 75 MHz.

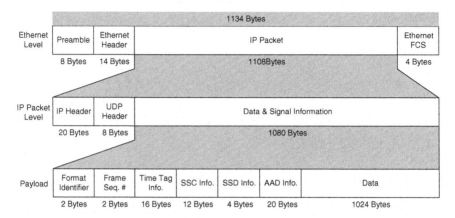

Fig. 7.15 SDDS packet format

- Accurate specification of the digitizer sampling clock. Not only is the nominal frequency specified in every packet, but the rate of change of that frequency is specified as well. These two pieces of information can be used to very accurately recreate the digitizer clock.
- Two timestamps are generated

  1. Specify the UTC time for the first sample in the packet
  2. Specify the sample within this packet that occurred on a 1 ms boundary. The 1 ms time tag is used to create time tags for even older standards (Datolite and SDN)

- Two low-rate data streams are interleaved with the primary sampled data.

  1. A synchronous serial data stream (SSC) consists of one data bit that transitions infrequently (no more than once per packet) to mark a specific sample. The sample number at the transition is sent with every packet. The significance of the marks is not specified (e.g., could indicate that the antenna was moved).
  2. A 19.2 kbps asynchronous data stream (AAD) is used to provide 'side' information about the main data stream (compare to the context packets described in Sect. 7.4.1). The meaning of the data stream is not specified.

Simple packet-level error correction is explicitly supported. Every 32nd packet can contain parity data over the preceding 32 packets. This allows the system to correct any single packet that is in error out of the 32 packets.

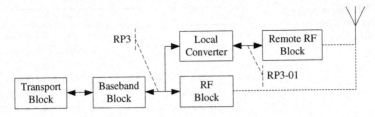

Fig. 7.16 Base station architecture for OBSAI

7.4.4 Open Base Station Architecture Initiative

The OBSAI family of standards was developed to address interoperability between subsystems in a wireless base station. A modular base station architecture assumed by OBSAI is shown in Fig. 7.16. The interfaces between subsystems are known as reference points (RPs). Digital IF interface between the DSP and mixed-signal/RF is described in RP3 and RP3-01 [119]. RP3-01 extends RP3 by adding a way to send control messages from the control block (RP1) to the remote RF block. Unlike digRF, which defines yet another standard for the physical layer, OBSAI leverages existing standards.[13] At the highest line rate, the link can support one signal with bandwidth up to 100 MHz or multiple narrower band signals [120].

Since OBSAI was developed to support commercial base stations, the standard has explicit support for popular wireless standards: GSM/EDGE, WCDMA, LTE, WiMAX. Packet formats are described to transport digital IF samples for each standard. Every packet contains the type field and a timestamp. Timestamps are required to make packet-based digital IF transfer work. For OBSAI, the timestamp is interpreted differently for uplink (base station receiver) and downlink (base station transmitter):

- In the uplink direction, a timestamp identifies the time instant when the last sample of the message was available at the output of the channelizer block (down-converter, FIR filter).
- In the downlink direction, a timestamp defines the time instant when the first sample of the payload must be inserted into the modulator (up-converter, FIR filter) (see also discussion in Sect. 6.1 and Fig. 6.3).

Interpretation of the timestamp also depends on the packet type (i.e. wireless standard). For example, a WCDMA timestamp refers to the index of the chip in a timeslot. For WiMAX, timestamp refers to the sample number in a frame.

The packet payload also depends on the packet type. For example, downlink WCDMA samples use 16 bits to encode I and Q components of each chip, sampled once per chip, while the uplink uses 8-bit values sampled twice per chip. The

---

[13] OBSAI goes one step further than digRF and specifies even the mechanical connectors, but that part of the standard is not relevant to this section.

Table 7.4 Partitioning of
DSP and RF subsystems in
the CPRI documentation

Downlink	Uplink
Pulse shaping	Matched filtering
D/A conversion	A/D conversion
Upconversion	Downconversion
On/Off control of each carrier	Automatic gain control
Carrier multiplexing	Carrier de-multiplexing
Power amplification	Low-noise amplification
Antenna supervision	
RF filtering	RF filtering
Measurements	

dozen or so different formats are clearly not sufficient for a general SDR. It is
interesting to note that all formats specify baseband sampled data (I/Q) rather than
IF sampled. Thus, technically, OBSAI cannot be considered a digital IF standard—
only a digital baseband standard.

A large part of the standard is devoted to specifying frequency stability, frequency accuracy, and all the delays in the system (including delays through RF
cables). Very tight tolerances are required to support multi-antenna techniques.
This level of specification is not provided in VITA 49 standard, but is absolutely
necessary to support MIMO operation.

7.4.5 Common Public Radio Interface

The CPRI standard was developed by a consortium of competitors[14] to the consortium that developed OBSAI. CPRI also addresses the cellular base station
market and supports most of the same wireless standards as OBSAI [121]. CPRI
was initially developed to support only the WCDMA wireless standard. Since then
it has been expanded to support WiMAX and LTE as well. True to its single-
purpose roots, CPRI specifies not only the format of the data exchanged between
DSP (known as REC in CPRI documentation) and RF (known as RE) blocks, but
even the exact partitioning of the radio. As shown in Table 7.4, the RF block starts
with pulse shaping. This level of specificity makes CPRI unsuitable for a generic
SDR.

Similar to OBSAI, all of the data formats assume baseband (IQ) sampling.
However, different sample precision is supported (4–20 bits for uplink, 8–20 for
downlink) and the oversampling ratio can be either 1 or 2. The additional sample-
level flexibility requires more complicated packet formats than OBSAI. The data is
sent over a high-speed serial link with line rates up to 10 Gbps in multiples of

---

[14] Ericsson, Huawei, NEC, Nokia Siemens Networks and Alcatel-Lucent.

614.4 Mbps. The fundamental line rate was chosen to work well with WCDMA standard, but requires stuffing of dummy bits to support WiMAX.

Control and management (C&M) data is time interleaved between the IQ data packets. Two different layer 2 protocols for C&M data—a subset of High-level Data Link Control (HDLC) and Ethernet—are supported by CPRI. These additional control and management data are time multiplexed with the IQ data. Finally, additional time slots are available for the transfer of any type of vendor-specific information.

Chapter 8
Software-Centric SDR Platforms

GPPs have become fast enough to implement all of the DSP for practical radios. As discussed in Sect. 5.1, an all-software radio is considered the holy grail of SDR development [122]. Presently, the two most popular open-source SDR platforms are GNURadio and OSSIE. In this section, we will explore these platforms in some detail.

8.1 GNURadio

GNU Radio (GR) is an open-source framework for designing SDRs. The major components of GR are:

1. A framework for an arbitrary signal processing block that can be connected to one or more other blocks.
2. A scheduler responsible for activating each processing block and managing data transfer between blocks.
3. C++ and Python infrastructures to build up a flowgraph from multiple blocks and attach the flowgraph to the scheduler.
4. A relatively rich set of frequently used DSP blocks such as filters, tracking loops, etc.
5. A graphical user interface (GUI) that allows the user to create GR designs by drag-and-dropping blocks and drawing connections between them (see Fig. 8.1).
6. An interface to a commercial hardware front end. The front end hardware provides the interface between the GPP and the physical world. It contains mixed-signal (ADC, DAC) and radio frequency (up/downconverter) components (see Sect. 8.4.2).

GR waveforms are designed using datapath architecture (see Sect. 6.4.2), which is ideally suited for processing streaming data because data is transferred between blocks via streams. A packet-based interface using message queues is available, but is not intended for transferring real-time data samples.

E. Grayver, *Implementing Software Defined Radio*, DOI: 10.1007/978-1-4419-9332-8_8, © Springer Science+Business Media New York 2013

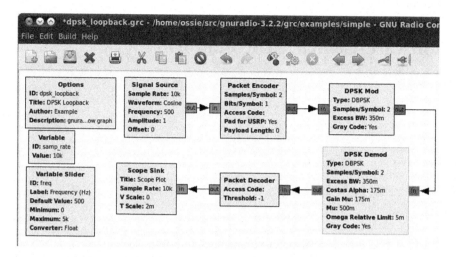

Fig. 8.1 GNURadio Companion graphical user interface (GRC)

Most of the signal processing blocks are developed in C++. The blocks can be interconnected, and complete GR applications can be created using either C++ or Python. Python is an interpreted[1] (rather than a compiled) language, and is generally considered to be easier for development and debugging. All the visualization blocks (e.g., scope, spectrum) provided with GR rely on the Python development flow. True to its open-source roots, GR is best supported on Linux (and MacOS), with less support for Windows.

8.1.1 Signal Processing Blocks

A block is the fundamental unit of a waveform developed using GR. A block may be as simple as a multiplier (a.k.a. gain), or as complex as an FEC decoder. A block is characterized by its inputs, outputs, and the ratio (r) of the input rate to the output rate.

A block can have an arbitrary number of input and output ports, with the actual number determined once the flowgraph is built.[2] Input and output ports can each be of a different data type (e.g., floating point, complex value, etc.) with different sample rates. Three special types of blocks are identified:

---

[1] An open-source software tool called SWIG is makes blocks developed in C++ available in Python.

[2] For example, an 'adder' block can be declared with an arbitrary number of inputs and one output. The block can then be used to sum up multiple data streams, with the number of streams determined at runtime.

Table 8.1 Characters that describe data types for GR blocks

Character	Type
b	Byte
s	Short (2 bytes)
i	Integer
f	Single precision floating point
c	Complex floating point
v	Modifies the subsequent character to indicate a vector of that type

1. *Source* blocks have no inputs and one or more outputs
2. *Sink* blocks have no outputs and one or more inputs
3. *Hierarchical* blocks contain one or more basic blocks.

All GR blocks must be derived from one of these fundamental types:

1. gr_block allows arbitrary *r*, and *r* can change over time
2. gr_sync_block requires *r=1* on all ports
3. gr_sync_decimator requires *r=N* on all ports, where *N* is an integer
4. gr_sync_interpolator requires *r=1/N* on all ports

The block name should end in a suffix that indicates the data types it consumes and generates.[3] The suffix consists of one, two, or three characters. The first character denotes the input data type, the second character denotes the output data type, and the third character denotes the coefficient data type for filters. Characters for standard data types are listed in Table 8.1. A few examples are given in Table 8.2.

The structure of a gr_block is shown in Figure 8.2. The key component supplied by GR is the output buffer. The output buffer is a thread-safe single-writer, multiple-reader circular buffer. The user-defined work(...) function processes data from the input ports and puts the output into the buffer(s). Data are passed to the work function as an array of abstract items. Note that before the input data can be processed, it must be cast into the appropriate data type (e.g., integer).

Each item can be annotated with an unlimited number of metadata 'tags.'[4] A tag is defined by a tuple {name, value, origin}, where the name is a string describing the tag, and value can be of any data type.[5] The origin field is optional and is meant to convey which block generated the tag. For example, the seventh item on input N in Figure 8.2 is annotated with the time (0.0123) that item was captured. Three tag propagation options are available:

---

[3] The naming convention is not enforced and some blocks do not follow it.

[4] Only a few items are expected to be tagged since tags are processed much slower than the items themselves.

[5] GR defines an abstract polymorphic type, pmt_t. This type acts as a container for arbitrary data types (e.g. doubles, floats, structs, arrays, etc.).

Table 8.2 Samples of the GR block naming convention

Block name	I/O description
gr_sig_source_s	A signal source that outputs 'shorts.' Note that a source block has only one type associated with it.
gr_vector_sink_b	Converts input 'byte' stream to a vector of bytes. Note that a sink block has only one type associated with it.
gr_add_ff	Adds two or more streams of floats to generate a float output
gr_add_const_vcc	Adds a complex valued constant vector to an input vector of complex values.
gr_fft_vfc	Computes an FFT of an input vector of floats and output a vector of complex values.
gr_fir_filter_ccf	A filter with floating point coefficients processes complex input and generates a complex output.

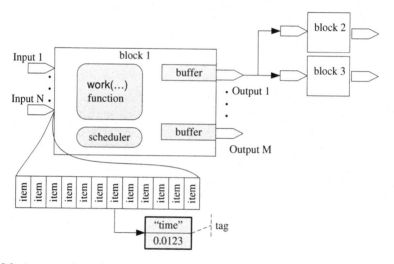

Fig. 8.2 Anatomy of a gr_block

1. Copy tags from i^{th} input to i^{th} output. This is the default option and is only applicable if the number of inputs is equal to the number of outputs.
2. Copy tags from all inputs to all outputs.
3. Do not copy any tags automatically. The user can then write code to selectively propagate and optionally modify tags.

Automatic tag propagation correctly handles the input/output ratio. For example a 1:10 interpolator propagates an input tag for item n to output item $n \times 10$.

Once all downstream blocks have completed processing a range of data, the write pointer in the circular buffer is advanced. This simple and elegant architecture allows a programmer to concentrate on the signal processing and not worry about transferring data between blocks.

In addition to items and tags, a block can send and receive messages. Messages are similar to tags in which they can contain arbitrary data. Unlike tags, messages are asynchronous, i.e., messages are not tied to items. Messages are typically used to exchange data between a top-level GUI or control application and the blocks.[6]

To create a new GR block, the user must be familiar with C++ and optionally with Python. The block is developed in C++ and must define only one function, work(), that implements the actual signal processing.[7]

The work function consumes the bulk of computational resources and must therefore be efficient to meet real-time processing throughput. Some frequently used blocks (e.g. filters, FFT) provided with the GR framework are optimized to take advantage of special instructions available on modern processors (i.e., hand-coded in assembly to utilize SIMD instructions). However, most of the blocks are written in standard C++ and rely on the compiler for any optimization. In the author's experience, simply using a processor-specific optimizing compiler such as the one from Intel increased the throughput for a simple radio by a factor of two. Even greater improvement can be achieved by using highly optimized libraries provided by the processor vendor (e.g., IPP and MKL from Intel [123], ACML from AMD, and others). Modifying the work function of most GR blocks to take advantage of the IPP library increased the throughput by a factor of four[8] (see Sect. 8.1.4).

8.1.2 Scheduler

The current GR scheduler is multithreaded[9]—each block runs in a separate thread, with its own mini scheduler [124]. The overhead due to switching between many threads and handling the semaphore locks on the data buffers is negligible as long as many samples are processed every time a block is activated.[10] The scheduler keeps track of how much space is available in the block's output buffers and how

---

[6] The top-level GUI or control application can also call member functions of the blocks directly. Messages provide a higher level interface that is guaranteed to be thread-safe. Messages also make it easy to control radios that execute on multiple computers.

[7] Many blocks also define functions to set parameters at runtime (e.g., set the gain on a 'multiply' block).

[8] The performance improvement is strongly dependent on the coding style. GR developers are actively working on integrating processor-specific optimizations into the GR baseline. The latest version of GR provides a library of hand-coded assembly functions called VOLK to take advantage of modern processors [287]. Users can expect baseline GR performance to improve, making it closer to what is achievable with an optimizing compiler and vendor libraries.

[9] Older versions of GR used a single threaded scheduler.

[10] This constraint is not always easy to achieve. Multiple blocks generating and consuming data at different rates cause 'fragmentation.' Fragmentation means that the scheduler calls the work function with only a few samples. Larger sample sets can be enforced by setting a minimum number of samples a block can accept. The problem is exacerbated if feedback loops exist between blocks. The author has observed scenarios where 90 % of the CPU time was spent in the scheduler itself rather than in the work functions. These anomalous conditions are difficult to identify and debug.

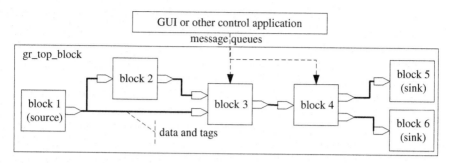

Fig. 8.3 GR top-level block and control application

much data are available from the upstream blocks. The scheduler then tries to maximize the amount of data processed at once. The block activation schedule is a function of space in the buffers and the sample rate change introduced by each block. This form of flow control is known as event-driven, where an event from an upstream block initiates processing in downstream block(s). For example:

- Block 1 has 1,000 samples in its output buffer
- Block 2 has 500 free samples in its output buffer
- Block 2 is an interpolator with $r = 4$.

Block 1's scheduler wakes up block 2's scheduler to tell that data are available. Block 2's scheduler calls its work() to process min(1,000, 500/4)=125 samples from block 1. The scheduler also takes care of passing messages between blocks using message queues. The scheduler is automatically created when the user derives a top-level block from a gr_top_block class, as shown in Fig. 8.3.

8.1.3 Basic GR Development Flow

GR is installed from a binary package available for Linux or Windows, or built from source code. Documentation is very sparse, even for an open-source project, and the developer has to be prepared to learn by example. A number of example designs are provided in the source code to demonstrate many fundamental features, mostly for Python and a few for C++.[11] One of three development flows must be selected: (1) Python, (2) C++, (3) Graphical (GRC). The GRC flow has the lowest learning curve and is recommended for relatively simple designs. Python flow is recommended if the developer is familiar with that language. C++ flow can be used if the developers are not familiar with Python, or need a completely stand-alone application that does not rely on Python.

---

[11] More full-featured examples are available at the GR archive network (GRAN) at https://www.cgran.org.

Starting with a system block diagram of the desired radio, the developer must identify which blocks[12] are already provided with baseline GR distribution. Documentation for each block is entirely in the C++ header file for that block and varies from nonexistent to comprehensive. Blocks that are not provided in the distribution have to be created by either writing C++ code or by combining available blocks. While it is possible to build up blocks from many primitive blocks (e.g. add, multiply, mux, etc.), the resulting design is very difficult to read and typically results in lower throughput. For any relatively complex radio, the developer will have to write one or more blocks from scratch.

File source and sink blocks are used extensively during debugging. These blocks capture intermediate and final signals to disk for post processing. For example, a single block can be debugged by creating a flowgraph consisting of a file_source, block_under_test, and file_sink.[13] Real-time debugging is supported by a combination of graphical output blocks (scope, spectrum) and file sinks.

In general, it is difficult to estimate the maximum throughput based on the flowgraph. GR documentation does not provide any benchmark data. If the processor cannot keep up with real-time throughput, the hardware interface blocks will output under/overflow warnings. Classical software performance improvement techniques can be applied to increase throughput. A high-quality profiler (e.g. VTune or gprof) is invaluable in identifying the bottleneck blocks.[14]

8.1.4 Case Study: Low Cost Receiver for Weather Satellites

The Geostationary Operational Environmental Satellite R-series (GOES-R) satellite [125] is being developed by National Oceanic and Atmospheric Administration (NOAA) to provide the next generation of weather data. One of the missions of GOES-R is to provide support for future versions of the Emergency Managers Weather Information Network (EMWIN) and Low-Rate Information Transmission (LRIT) signals [126, 127]. The current signal bit rates are below 300 Kbps. GOES-R increases the rates to 1 Mbps. The cost of upgrading ground stations, especially for developing world countries, is a major concern for NOAA. NOAA developed an SDR to demonstrate the feasibility of a low-cost terminal that can be used to receive both GOES-R and legacy signals [64].

GNURadio was selected for this effort since it is open source and does not increase the cost of the receiver. A number of new blocks were designed to demodulate and decode EMWIN and LRIT satellite signals. All the new blocks

---

[12] A mostly complete list of fundamental blocks is available at http://gnuradio.org/doc/doxygen/modules.html.

[13] GR development guidelines suggest using automatic regression testing for every new block.

[14] Common sense is usually insufficient. The author came across a design where the noise generation block took significantly more computational resources than a Viterbi decoder. Also see footnote 10 in this chapter.

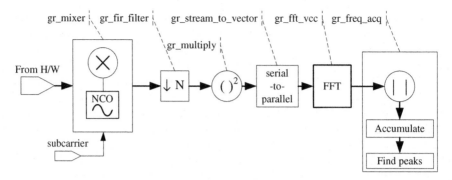

Fig. 8.4 Block diagram of the frequency acquisition engine

follow GR block architecture design and can be shared with other GR developers. Many blocks from baseline GR distribution were redesigned to use Intel Integrated Performance Primitives (Intel IPP) for improved performance [123].

The only incremental cost (versus an existing ground station) is a front end capable of digitizing the new, wider band signals. A custom digitizer designed for this effort, described in Sect. 8.4.2.1, can be built for <$100.

The next sections describe some of the signal processing required to demodulate three signals of interest: LRIT-BPSK, EMWIN-OQPSK, EMWIN-FSK.

8.1.4.1 Frequency Acquisition

The constituent blocks of the acquisition engine are shown in Fig. 8.4. Each block in the figure corresponds to a GR block executing in a separate thread. All the blocks except for the `gr_freq_acq` are part of the standard GR distribution. Frequency is acquired by finding a peak in the averaged fast Fourier transforms (FFTs) of the signal. The search window (frequency uncertainty), resolution, and amount of averaging are all runtime programmable. An Nth-power detector is used to remove modulation from the subcarrier. Peak search is run once the desired number FFTs have been averaged. The acquisition engine returns locations and powers of detected peaks. Since the signal can fall between FFT bins, experiments show that considering multiple adjacent bins improves acquisition accuracy. The acquisition result is used to adjust the frequency of the downconverter in the digitizer.

8.1.4.2 BPSK and OQPSK Demodulation

A conventional receiver with separate phase and timing tracking loops was implemented to demodulate BPSK and OQPSK signals (Fig. 8.5). A Costas phase tracking loop removes any frequency offset remaining after acquisition and aligns the phase. A Mueller–Muller timing tracking loop [128] aligns the sampling time

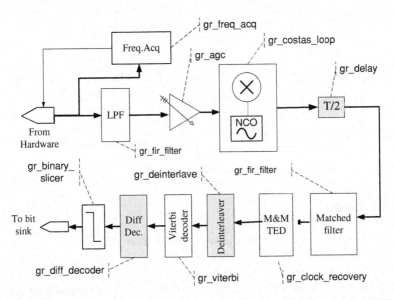

Fig. 8.5 Block diagram of the BPSK/OQPSK demodulator (shaded blocks only for QPSK)

to the peak of every symbol. The symbols are then processed by a Viterbi decoder. Designing a software Viterbi decoder to handle the data rate of 900 kbps was challenging. The decoder supports any constraint length, rate, and polynomial. Since the specifications of the GOES-R broadcast signal are not completely defined, the advantage of having a flexible decoder outweighs the potential for higher throughput from a fixed decoder. The initial design was able to maintain a throughput of 300 kbps. The algorithm was vectorized to take advantage of the IPP functions and throughput increased to 1.9 Mbps.

8.1.4.3 FSK Demodulation

Some of the data from older weather satellites is transmitted with FSK modulation. FSK is very robust and may be available even when signals described in the previous section cannot be received. The FSK signal encodes an *asynchronous* waveform that closely mimics the serial interface used in dialup modems. A block diagram of the FSK demodulator is shown in Fig. 8.6. The symbol decision metrics are created by correlating the received signal with two locally generated tones, corresponding to the two transmitted tones. The magnitude of the output of the correlator, implemented as an integrate-and-dump block, is sampled at symbol boundaries, and a hard decision is made. The FSK demodulator does not require any frequency tracking since the front end is assumed to compensate for large frequency offsets. Small frequency offsets have a negligible effect on the overall performance of the system. The FSK symbol timing is acquired by looking for a '01' sequence transition, and then tracked using the same algorithm as for BPSK.

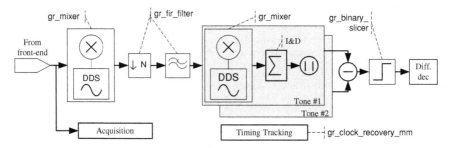

Fig. 8.6 Block diagram of the FSK demodulator

8.1.4.4 Results

The implementation of a complete receiver for three different live signals demonstrates that GR is suitable for deployment and is not just for experimentation. The lost cost, very flexible receiver demonstrates some of the advantages of SDR. The front end digitizer supports legacy signals as well as proposed signals, and has enough bandwidth for future signals as well. The entire development effort took less than 2 man-years, which is significantly faster than implementing the same receiver(s) in an FPGA or ASIC.

8.2 Open-Source SCA Implementation: Embedded

Open-source SCA implementation—embedded (OSSIE) is an open-source platform for developing SCA-compliant (see Sect. 7.1.1) radios.[15] At first glance it is somewhat similar to GNURadio in that it also allows a developer to create and interconnect signal processing blocks. While GR was developed from the ground up by a team of dedicated amateurs for quick-and-dirty experimentation on a budget, OSSIE complies with a standard developed by the military at a cost of billions. GR assumes it is running on a Linux variant and has direct access to the file system and hardware. SCA waveforms run in a well-specified operating environment, with the file system and hardware abstracted by an API. Blocks in GR are interconnected using explicit Python or C++ commands and samples are transferred using custom circular buffers. OSSIE uses XML files to define the connectivity, and the actual connections are accomplished using CORBA services. More importantly, samples are transferred over the ORB, using TCP/IP. In particular, the ORB allows different blocks to execute on different machines, while

---

[15] In this section, SCA and OSSIE will be used interchangeably.

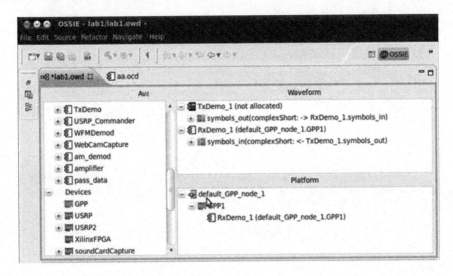

Fig. 8.7 OSSIE GUI for defining waveforms

GR flowgraphs are meant to execute on a single machine.[16] Since ORB is a general purpose software bus and is not specific to SDR, the concept of a 'scheduler' does not really apply to an SCA waveform. Samples are transferred over the ORB and basic flow control is used to start/stop transfers.

The main goal of GR is providing the signal processing framework, while the main goal of OSSIE is providing an SCA operating environment.

These differences and similarities have led to significant confusion and some developers still regard GR as a competitor to OSSIE [129]. In fact, the OSSIE team has recently demonstrated an SCA-compliant waveform that relies on GR for signal processing. A GR flowgraph is encapsulated inside an SCA component [130].

It is quite feasible to develop a complete radio in OSSIE, although the set of signal processing blocks provided with the baseline distribution is significantly smaller than for GR.

The OSSIE distribution includes:

- A GUI (based on the popular Eclipse platform) to select and interconnect blocks (see Fig. 8.7). Note that the connections between blocks are indicated by icons instead of lines (TxDemo_1 is connected to RxDemo_1).
- A GUI to define new blocks. This GUI creates the skeleton C++ files that must then be modified by the developer to add the desired signal processing code.[17]

---

[16] It is relatively easy to build an application in GR that runs on multiple computers. For example, the UDP source/sink blocks can be used to transfer samples between machines. However, this approach is ad hoc, versus the well-defined ORB architecture.

[17] Only the: ProcessData() function is modified to implement the desired signal processing. This is very similar to modifying the: work() function in a GR block.

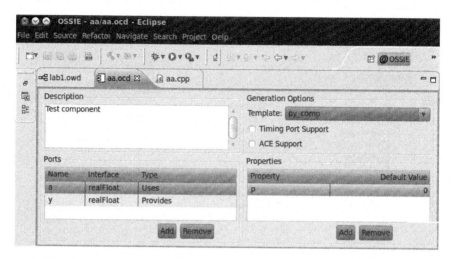

Fig. 8.8 OSSIE GUI for creating new components

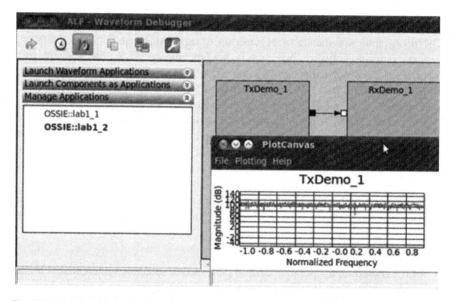

Fig. 8.9 OSSIE GUI for viewing and debugging application

For example, a component shown in Fig. 8.8 has an output (uses) port named a, an input (provides) port named y, and a configurable parameter called p.

- A GUI to examine and debug waveforms (applications). This GUI allows the developer to monitor (scope, spectrum) interblock signals. Monitors can be added at runtime (unlike GR, which requires all the monitors to be defined before the flowgraph is started). The debug view of the application defined in Fig. 8.7 is shown in Fig. 8.9.

- A fully SCA-compliant CORBA based on the open-source omniORB.
- A set of tutorials and laboratory exercises. OSSIE is distributed as a ready-to-run virtual machine. A Windows user can download the distribution and execute it without having to install Linux. Simple installation and tutorials make it somewhat easier to get started with than GNURadio.

8.3   Other All-Software Radio Frameworks

GNURadio and OSSIE are by far the most popular open-source SDR development platforms.

There are a few commercial products that compete with OSSIE. The popular Spectra system from Prismtech [131] provides a complete environment for developing SCA-compliant waveforms. It comes with a high-performance CORBA implementation and is geared toward military-grade waveform development. Prismtech offers a complete solution, including a hardware front end tightly integrated with the software. One of the unique features of the Spectra framework is support for *native* FPGA implementation of a subset of CORBA features (see Sect. 7.1.1.2). This FPGA-based CORBA allows components to be deployed on FPGAs without the use of proxies or other hardware abstraction layers (HALs) that impose additional memory and processing overhead. Other SCA framework vendors include L3 Communications, BAE Systems, Harris, and CRC [132]. CRC provides both a commercial and open-source SCA framework.

There is no direct commercial equivalent to GNURadio.[18] Simulink from Mathworks can be used to create flowgraphs that implement SDR. In fact, Simulink supports the same hardware front end as GR. However, Simulink is a more general purpose environment and is much slower than GR for implementing even moderately sophisticated waveforms. The Microsoft Sora framework is a quasi-commercial competitor to GNURadio and will be discussed in the next section.

8.3.1   Microsoft Research Software Radio (Sora)

Many corporations have robust in-house SDR research efforts. Some have decided to leverage GNURadio, while others are building frameworks from scratch.

---

[18] The venerable X-Midas (multiuser interactive digital analysis system for X-windows) framework is a quasi-open-source alternative to GNURadio. X-Midas has been in development since the 1970s and is used exclusively by the intelligence community. It is only available to qualified users from the government. X-Midas provides most of the functionality available in GR, and a simple wrapper can be created to make GR blocks execute 'as-is' in X-Midas. A parallel development from the X-Midas community is available to the general public [288].

Microsoft has released some information about their in-house SDR framework called Sora [133]. Unlike GNURadio, Sora is only available to academic institutions, is not licensed for commercial use, and is not open source. Similar to GNURadio, Sora consists of a library of prebuilt signal processing modules and an analog front end. Unlike GR, which focuses on standard and portable code, Sora aims to achieve the highest possible throughput by taking advantage of every available optimization. The key optimization technique used by Sora is mapping complex mathematical operations into precomputed lookup tables. Once the tables are initialized, each operation is very fast as long as the tables fit within the processor cache. One example is a soft demapper (also known as a slicer), which converts the received complex symbols into soft decisions suitable for FEC decoding. Performance-critical signal processing code is optimized to use processor-specific single instruction multiple data (SIMD) instructions. Since the SIMD registers on Intel processors are 128-bit wide while most PHY algorithms require only 8 or 16-bit fixed-point operations, one SIMD instruction can perform 8 or 16 simultaneous calculations!

Just like GR, Sora uses a scheduler to activate each signal processing block and passes data between blocks. However, unlike GR, the Sora scheduler uses a static schedule optimized for the particular flowgraph. The number of circular buffers between the blocks and the number of threads is therefore much smaller than for an equivalent GR flowgraph.[19] For optimal processor utilization, the number of threads should be the same as the number of cores, as long as each thread can be guaranteed to use all of the core's resources. Threads are mapped to dedicated cores to minimize cache thrashing. The Sora architecture reduces performance penalty when processing small blocks of data. GR performance plummets when the block size falls below 512 bytes.

The results presented in [133] show that Sora can achieve about $10\times$ higher throughput than a similar GR implementation. However, FEC decoding is still a major bottleneck—even the relatively simple Viterbi decoder requires a dedicated core. Today's GPP is still not fast enough to decode powerful modern codes [134].

8.4 Front End for Software Radio

The SDR frameworks described in this chapter all require an interface to the 'real world.'[20] As discussed in Chaps. 2 and 9, all practical radios today require an RF and mixed-signal front end. Custom hardware is usually developed for radios intended for commercial or high volume. A front end is usually integrated with a GPP on a single circuit board to meet the mechanical requirements of the radio.

---

[19] A static scheduler also avoids data the fragmentation problem discussed in footnote 10 in this chapter.

[20] Unless the SDR is used for simulation rather than for actual communication.

Radios intended for research and development typically leverage COTS solutions. Describing commercially available products in a book is a losing proposition since they are invariably obsolete by the time the book is published. The hardware described in this section is meant to represent different classes of front ends available to SDR developers rather than endorse any particular solution. Some solutions provide only the mixed-signal front end and require additional hardware for the RF front end. Separating mixed-signal from the RF allows the user to mix-and-match products and select the best combination for each radio. However, recent availability of inexpensive wideband RF hardware has made integrated solutions possible.

8.4.1 Sound-Card Front Ends

The simplest and lowest cost mixed-signal solution is already available in all PC in the form of the sound card. A high-end sound card has A/D and D/A converters capable of digitizing ~ 40 kHz of bandwidth and can therefore support some narrowband waveforms (e.g. AM or FM radio, some amateur radio signals). Sound card-based SDRs are used primarily as educational tools [135, 136], but have also found practical application [137]. An RF front end is not required for basic experimentation, and two sound cards can be connected directly to each other.[21] The amateur radio community offers a few RF front ends designed to interface directly to a sound card [138] for under $200.

8.4.2 Universal Software Radio Peripheral

Introduction of the Universal Software Radio Peripheral (USRP) brought down the cost of SDR experimentation from tens of thousands to hundreds. The USRP is a commercial product initially developed by a one-man shop that was then acquired by National Instruments. USRP was released at about the same time as GNURadio and has been well integrated into that framework. In an unusual development model, the source code for USRP drivers and the mixed-signal circuit board are in the public domain, while RF circuit boards are proprietary. USRP is now also supported in Simulink [139] and LabView [140]. The first generation of USRPs relies on the USB2 interface and provides up to ≈ 30 MHz of instantaneous bandwidth for about $1000 (Fig. 8.10b). The second generation moved to gigabit Ethernet and provides up to 50 MHz of bandwidth for about $2000 (Fig. 8.10c).

---

[21] The signal could even be sent over the air, but an enormous antenna is required to efficiently transmit a 15 kHz carrier (wavelength $= c/15e3=20$ km). Practically, a smaller antenna would work at the cost of reduced SNR.

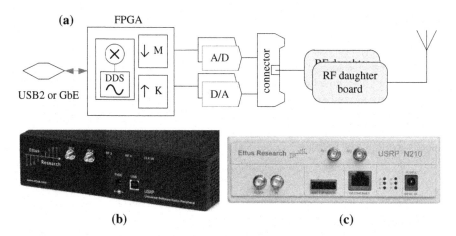

Fig. 8.10 Universal software-defined radio peripheral (USRP). **a** Block diagram. **b** First-generation (USRP1). **c** Second-generation (USRP2)

Different daughter boards are used to support a wide range of RF frequencies, with almost continuous coverage from DC to 6 GHz. Each USRP supports two receive and two transmit channels,[22] and multiple USRPs can be synchronized to provide more coherent channels. The USRP consists of an FPGA, 100 MHz-class A/D and D/As, RF daughter board(s), and an interface to the host computer (Fig. 8.10a). The FPGA is used to channelize the data and interface to the host port. The Ethernet interface on the USRP2 supports the VITA-49 standard discussed in Sect. 7.4.1.

The FPGA implements digital frequency shifting and rate change. The FPGA source code can be modified to implement more sophisticated processing. For example, multiple channels can be digitized in parallel and sent to different computers on the network. Some or even all[23] of the DSP for the SDR can be implemented on the FPGA to achieve higher throughput than is possible with a PC [141]. One version of the USRP2 integrates a GPP and DSP into the chassis and can be used to implement a completely stand-alone SDR.

8.4.2.1 Low-Cost USRP Clone

A clone of the USRP1 was developed to address the need for a low-cost receiver. The receiver was developed by The Aerospace Corporation as a proof of concept for the next-generation satellite weather station (see Sect. 8.1.4). A stripped-down USRP1 includes a single receive-only channel and an integrated RF front end. This device, shown in Fig. 8.11, has a bill of materials under $100.

---

[22] First-generation USRP provides four A/D and D/A channels, but RF daughter boards support only two.

[23] The FPGA is relatively small and is not suitable for complex DSP such as FEC decoding.

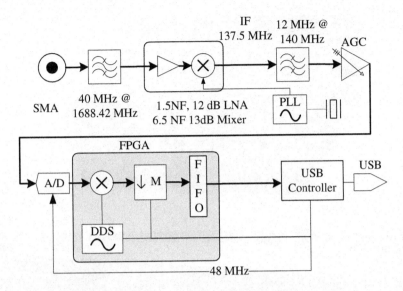

Fig. 8.11 Block diagram of a low-cost, receive-only USRP clone

Two front end interfaces were designed:

1. An IF interface for 140 MHz carrier frequency. This interface can be connected to an existing downconverter. The downconverters are available at all current ground stations to support legacy receivers.
2. An RF interface for 1.688 GHz carrier frequency. This interface is intended for new ground stations.

The input to the digitizer is assumed to have been received by a satellite dish and downconverted from radio frequency (RF) to an IF in the range of 140 ± 6 MHz. Figure 8.11 shows the overall composition of the receiver. Since 140 MHz is a popular commercial IF frequency, this allows us to use a low-power and inexpensive A/D and also provides for noise filtering using an inexpensive surface acoustic wave (SAW) input filter. To suppress out-of-band thermal noise, a 12 MHz wide SAW filter was used as a first element in the receiver chain. Insertion loss of the filter has negligible effect on the SNR since the signal has already been boosted by the LNA. Out-of-band noise is also greatly suppressed and will not be folded into the signal in the sampling process. The bandwidth of the filter was chosen to be larger than the bandwidth of the signal to accommodate the gamut of different weather satellite signals that may be of interest to the user. The SAW filter is followed by an automatic gain control (AGC) block which restores and keeps the signal level at an appropriate 2 V_{pp} differential level for the ADC. The AGC provides gain from -2.5 to $+45$ dB. The ADC used in the design has 12-bit resolution and a sampling rate of 48 MHz. The 140 MHz input signal is *subsampled* at 48 MHz. This technique allows us to relax the front end requirements for the digitizer since high frequency clocks and filters would increase the

(a)

(b)

Fig. 8.12 Network-based RF front ends

cost of the board. To minimize the overall cost of the digitizer, a single clock source is used on the board. As described previously, the signal is downconverted and decimated in the FPGA prior to transmission over the USB interface.

Components used in the USRP are too expensive to achieve NOAA's goal to build a receiver that could be manufactured for around $100. In order to reduce costs, dual ADC/DAC was replaced with a single ADC. The clocking network was overhauled so that only a single crystal is used instead of two crystals and a phased-lock loop (PLL). This crystal provides a clock reference for everything on the board while using the PLL inside the USB controller. A single receiver design fits on a smaller FPGA than the one on the USRP, reducing its cost by half. The components of the current board cost around $80, and even less if mass production cost reduction is considered.

The complete design consumes about 1.5 Watts and is powered entirely over the USB port. The board is enclosed in a metal box which reduces electromagnetic interference (EMI) emissions. The device also includes a set of LEDs to indicate the power of the input signal. The user can use the LED measurement as a feedback system to measure the incoming signal amplitude while adjusting the position of the satellite dish. This board features SPI and I^{2}C interfaces for internal control. All of these functions can be accessed through the general software interface, and most of the internal devices can be controlled in the same way.

8.4.3 SDR Front Ends for Navigation Applications

A few relatively low-cost RFFEs are available for developers working in the navigation bands (L-band). The low-cost SiGe front end [142] was made popular by an excellent book on software-defined navigation receivers [143]. NSL Stereo supports both navigation and non-navigation bands [144]. These and similar devices are suitable for implementing SDR receivers, but do not provide transmit capability.

8.4.4 Network-Based Front Ends

USRP is the most popular and by far the lowest cost general purpose SDR front end. From its inception, the USRP has been targeted at R&D rather than for deployment. Two representative front ends, shown in Fig. 8.12, are intended for field use and are available from thinkRF [114] and Rincon [145]. The thinkRF device includes both RF and mixed-signal subsystems, while the Rincon device requires an additional RF downconverter. Both devices provide only receiver functionality. The thinkRF device uses VITA-49 output, while Rincon uses the less common SDDS. These and other solutions [146, 147] are targeted at military and law enforcement users, and are about 5 times more expensive than a similarly configured USRP.

Chapter 9
Radio Frequency Front End Architectures

A wideband flexible radio frequency front end (RFFE) is needed to implement an SDR. As discussed in Chap. 2, the RF front end is the most challenging problem in a truly universal radio. In this chapter we consider some high-level tradeoffs in existing RF front end architectures and how they apply to SDR. The tradeoffs for receivers and transmitters are quite different. The transmitter is typically simpler to implement and will be considered first. As a point of reference, a modern cellular telephone RFFE, shown in Fig. 9.1, has dedicated circuitry for each of the supported bands.

9.1 Transmitter RF Architectures

A transmitter must generate an electromagnetic wave at sufficient power to ensure sufficient SNR at the receiver. The low-level signals generated inside the radio are amplified by a power amplifier (PA). The PA usually consumes more power than any other block in the radio. A PA is often a discrete chip since it requires a different fabrication processes than low-power components. The input to the PA should contain only the desired signal because it is wasteful to amplify signals and then filter them out. PA efficiency is defined as the ratio of output power to consumed power. Narrowband amplifiers are more efficient than wideband ones, but are not suitable for SDR since they restrict the range of supported carrier frequencies. Advanced signal processing can be applied to improve efficiency by operating the amplifier closer to its limit [149].

The next few subsections discuss a few techniques for converting digital samples to an RF waveform. These techniques are applicable to a wide range of waveforms and therefore suitable for SDR.[1]

---

[1] Techniques applicable to only one class of waveforms (e.g., constant envelope) will not be discussed. For a more detailed discussion see [222].

E. Grayver, *Implementing Software Defined Radio*,
DOI: 10.1007/978-1-4419-9332-8_9, © Springer Science+Business Media New York 2013

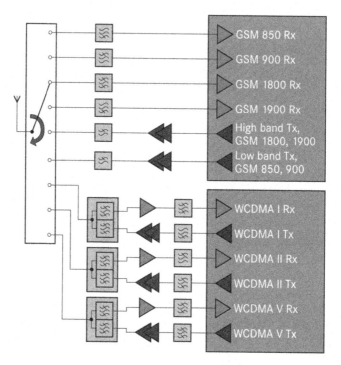

Fig. 9.1 RFFE interface for a multi-mode cellular phone [148]

9.1.1 Direct RF Synthesis

The basic purpose of the transmitter RFFE is to take the digital samples computed in the baseband subsystem and convert them to an RF waveform at the desired carrier frequency. The RF waveform is then converted to an electromagnetic wave at the antenna. The simplest transmitter RFFE is based on direct RF synthesis as shown in Fig. 9.2. The DSP subsection implements a digital frequency synthesizer (DDFS) with an effective[2] clock rate at least twice the desired *carrier frequency*, F_c. The DAC sample rate, F_s, must be[3] greater than $2 \times F_c$. The spectrum of the generated signal is shown in Fig. 9.2. Aliases of the desired signal occur at multiples of the sampling frequency, and must be filtered out before the power amplifier (PA). Note that as F_c approaches $F_s/2$, the separation between the desired signal and the alias decreases, requiring a sharper filter. Practical constraints on the

---

[2] The actual clock rate can be much lower if multiple parallel paths are used.

[3] This is not strictly the case since aliases of the primary signal in the higher Nyquist regions can be used [269]. However, practical concerns such as output power and passband flatness limit the use of Nyquist regions higher than 3.

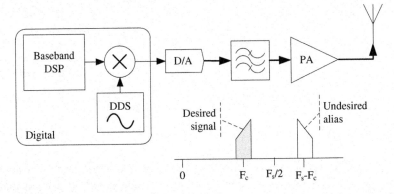

Fig. 9.2 Direct RF synthesis for the transmitter

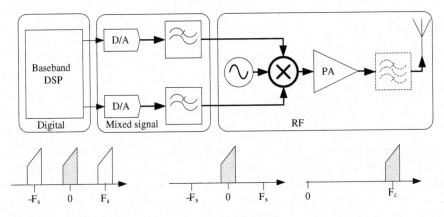

Fig. 9.3 Zero-IF upconversion with an optional (*dashed-line*) output filter

filter sharpness limit the maximum supported F_c to well below $F_s/2$. The advantages of direct RF synthesis are:

- Reduced number of components, leading to a more compact implementation and potentially lower cost.
- Fewer sources of noise and distortion.
- Very agile frequency synthesis allows the radio to jump from one frequency to another with almost no delay. This feature is especially important for cognitive radios and some wideband frequency-hopped (see Sect. A.4.9) systems.

Direct RF synthesis is rarely used in today's radios due to the following disadvantages:

- High sample rate is required for the DAC. For example, a 2.4 GHz carrier requires at least 5 GHz sample rate. DACs at this sample rate do exist (see

Fig. 9.4 IQ mixer

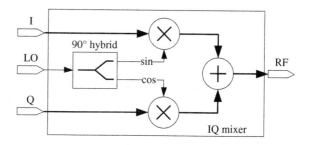

Sect. 10.2.4), but are expensive and draw a lot of power. Higher carrier frequencies cannot be generated directly because no adequate DACs are available.

Sect. 10.2.4), but are expensive and draw a lot of power. Higher carrier frequencies cannot be generated directly because no adequate DACs are available.
- High-speed DACs provide lower accuracy and dynamic range (fewer bits) than lower speed versions. High dynamic range is required for modern waveforms such as OFDM. Even more dynamic range is required if the transmitter supports multiple simultaneous users with very different power levels.

9.1.2 Zero-IF Upconversion

In a zero-IF architecture, an IQ mixer is used to translate the complex baseband analog signal to a desired carrier frequency. The name zero-IF refers to the fact that the input signal is centered on DC (0 Hz). Two DACs are used to generate the IQ components of a complex baseband signal, as shown in Fig. 9.3.

The mixer is usually a passive device. It internally creates a 90° shifted replica of the reference carrier (local oscillator, LO) and multiplies the I input by the LO and the Q input by the shifted replica. The sum of the products gives the baseband signal translated to the carrier frequency (Fig. 9.4). The ideal version of this component, known as an image-reject or quadrature-IF mixer, does not generate any additional spectral images and does not require filtering. However, practical broadband mixers create spectral images at each multiple of F_c and may require additional filtering before the PA.

The advantages of this architecture are:

- Baseband and DACs operate at a rate proportional to the bandwidth of the signal, not the carrier frequency. Lower sample rate results in reduced power consumption and possibly lower cost.
- Since two DACs are used to create the signal, each DAC must be sampled at only *1× the signal bandwidth* rather than 2×. Zero-IF is the only way to generate very wideband signals that exceed the bandwidth of a single DAC.
- Very high carrier frequencies can be supported with minimal increase in power consumption. IQ mixers are available for carrier frequencies from low MHz to well over 50 GHz, with a single device covering multiple octaves.

- Filtering can be very simple, with just two fixed-frequency lowpass filters at the outputs of the DACs to remove images above the DAC sampling frequency. Assuming a high-quality IQ mixer, no RF filtering is required.
- IQ mixers can be implemented in standard semiconductor processes and can be integrated with the DACs and even filters on a single chip.

The disadvantages are:

- The gain and phase for I and Q branches have to be matched very closely to achieve image rejection. A slight imbalance results in the appearance of a spectrally inverted image on top of the signal of interest.
- Any DC offset at the I or Q inputs results in LO leaking through to the RF port.
- Semiconductors introduce a lot of noise around DC due to a process known as 'flicker noise.' For narrowband signals this noise can significantly degrade overall SNR.
- A synthesizer is required to generate appropriate LO. The LO power level has to be fairly high (e.g. +10 dBm) to drive a passive mixer. The synthesizer can therefore burn a lot of power. An active mixer can be used, but also requires a nontrivial amount of power
- Two DACs are required instead of just one. This is a minor concern since dual-channel DAC chips are widely available.
- More circuit board space may be required for the two filters, synthesizer, and mixer. However, highly integrated (single chip) solutions exist.

The first two problems can be corrected in the digital baseband. If a feedback path is available to detect the presence of I/Q imbalance or LO feedthrough, the digital I and Q signals can be appropriately modified (e.g., by adding a small negative bias to the DAC input with a positive DC offset). The feedback may be always present for contiguous compensation, or may only exist during post-fabrication testing on the assembly line. In the latter case, RF output is measured using standard test equipment over the expected operating temperature range. Compensation coefficients are calculated and loaded into the DSP subsystem.

9.1.3 Direct-IF Upconversion

Direct-IF upconversion relies on the DAC to generate a signal at a low intermediate frequency (IF) and a regular mixer to translate the IF frequency to the desired carrier (Fig. 9.5). This architecture is an obvious extension of the direct-RF synthesis described above, but for higher carrier frequencies. The DAC is sampled at a rate proportional to the IF frequency rather than the carrier frequency.

The advantages are:

- Generating a signal at an IF frequency avoids 'flicker noise' that plagues zero-IF architectures.

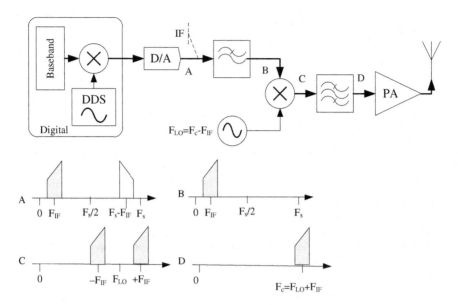

Fig. 9.5 Direct-IF upconversion

- DC offset in the DAC can be filtered out by the second bandpass filter, reducing the need for calibration.
- A simple mixer can be used[4] instead of a more complicated IQ mixer, possibly saving power.

  The disadvantages are:

- An RF bandpass filter is required to remove the image at F_{LO}–F_{IF}.[5] Depending on the desired carrier frequency range, the filter may need to be tunable.
- All of the disadvantages of zero-IF upconversion.

---

[4] A version of the IQ mixer, known as the Hartley Mixer, can be used to avoid the second bandpass filter. A Hartley mixer internally creates I and Q branches from the real-valued input signal, and then uses a standard IQ mixer.

[5] Alternatively, the image at $F_{LO} + F_{IF}$ can be removed. The remaining image is then spectrally inverted. Using high-side vs. low-side image provides a degree of freedom in the frequency plan of the radio. Interesting work reported in [292] proposes filter-less direct-IF upconversion based on polyphase combining.

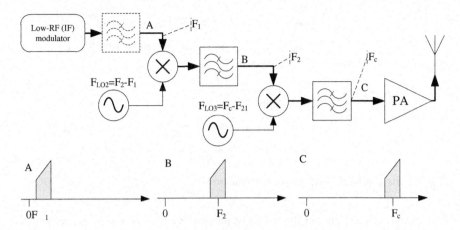

Fig. 9.6 Three-stage super hederodyne upconversion

9.1.4 Super Heterodyne Upconversion

Any of the previously described architectures can be extended by adding more RF mixers before the power amplifier (Fig. 9.6). The additional mixers are used to translate the signal to successively higher carrier frequencies. This approach is used when no single-stage components are available for the desired carrier frequency, or when the RF filter design becomes impractical.[6] For example, mixers for 100 GHz band do not work with baseband or low-IF inputs, but require inputs above one GHz. A 100 GHz transmitter would then use one of the architectures described above to upconvert a signal to, say, 2 GHz. The upconverted signal is then mixed with a 98 GHz LO to arrive at the desired 100 GHz carrier.

9.2 Receiver RF Front End Architectures

Receiver RFFE architectures are very similar to transmitter architectures. However, the tradeoffs are somewhat different. The main difficulty in designing wideband and flexible receiver front ends is the existence of adjacent channel interferers. Most (but not all) transmitters are only concerned with generating a single signal, usually of modest bandwidth. However, the receiver has to contend with *all* signals received by the antenna. As discussed in Chap. 2, it is not unreasonable for the desired signal to be adjacent to an interferer that is a billion times stronger.

---

[6] The Q-factor of the RF filter is roughly proportional to F_c/F_{IF} since the filter has to remove an image $2 \times F_{IF}$ away from the signal of interest, centered at F_c.

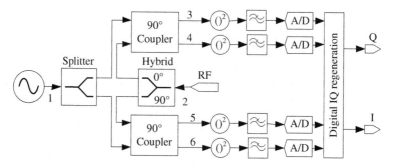

Fig. 9.7 Six-port RFFE receiver (ports are numbered)

A receiver must protect the signal of interest from interferers at every point in the receive path. This can be achieved with a combination of filtering (which reduces interference) and/or high dynamic range (which allows the DSP to remove interference). The tradeoff between dynamic range (required to linearly process the signal of interest and all interferers) and filtering (which removes the interferers) is fundamental to receiver RFFE design. High dynamic range requires all components in the receive chain to be very linear and the ADC to have high resolution (making them expensive and increasing power consumption). Filtering is problematic because tunable filters are not yet widely available (see 10.4).

Receiver architectures closely mirror the transmitter architectures (direct RF sampling, zero-IF downconversion, IF downconversion, and super heterodyne downconversion) will not be discussed for conciseness.

9.2.1 Six-Port Microwave Networks

Six-port microwave network is a technology developed in 1970s that has been recently repurposed for wideband, high frequency SDR RFFE [150, 151]. The idea behind six-port RFFE is to obtain multiple observations of the downconverted signal in the RF domain and then apply signal processing to extract the desired signal. It is somewhat similar to direct-IF conversion, but with more than two baseband signals. As shown in Fig. 9.7, four ADCs are used to digitize the magnitude (not the amplitude) of four downconverted versions of the signal. Each version is downconverted with a different phase. The additional observation allows DSP to compensate for mismatches in the RF components. The microwave network can be implemented in a compact and low-cost chip if the carrier frequency is high (>10 GHz). For lower carrier frequencies, the size of the six-port network is quite large. Obviously, the additional ADCs also contribute to the cost and power consumption. To the best of the author's knowledge, this technique has not been implemented in any commercial SDR RFFE.

Chapter 10
State-of-the-Art SDR Components

The state of the art in digital and RF components changes much faster than new editions of this book can hope to track. This section aims to capture the commercially available technology at the end of 2010 and is likely to be woefully out of date by 2015. However, even if technology has advanced, many of the trends and approaches described here will remain valid for a few more years.

The term 'state-of-the-art' must always be qualified by the specific market the art is addressing. Three broad market segments can be identified:

- Consumer-grade radios that are manufactured in high volumes and must be relatively low cost. For the purposes of this section, both cellular phones and military handheld radios are included in this segment.
- Research and development test equipment represents the high end of what is possible in the commercial marketplace. Cost is not entirely irrelevant, but is secondary to performance. These devices are hand built in low volume and extensively tuned and calibrated for optimal performance.
- Exotic, one-of-a-kind radios developed for the military represent the true state-of the art. Technology in these devices may never trickle down to consumers.

The cost of a single device increases exponentially in each market segment. A consumer-grade radio may cost of the order of $100; test equipment setup costs $100,000; and a radio on a military satellite can exceed $10,000,000.

10.1 SDR Using Test Equipment

The author has found that a good way to extrapolate what technology will be available for consumer-grade radios in 5–10 years is to look at the currently available test equipment. Test equipment includes instruments such as digital oscilloscopes, spectrum analyzers, and signal generators. Development of test

E. Grayver, *Implementing Software Defined Radio*,
DOI: 10.1007/978-1-4419-9332-8_10, © Springer Science+Business Media New York 2013

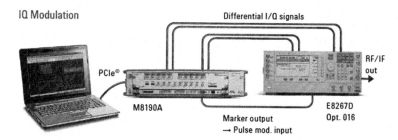

Fig. 10.1 Representative wideband SDR transmitter using Agilent test equipment [152]

equipment is market-driven and is targeted at current and upcoming wireless standards. It is not driven by the *desirements*[1] of SDR researchers.

10.1.1 Transmitter

A signal generator is essentially a very configurable transmitter, and is an excellent platform for an SDR. Two aspects of a signal generator must be considered: signal bandwidth and range of carrier frequencies. Since most current wireless standards target signals with less than 100 MHz of bandwidth, most signal generators also do not support higher bandwidth. The most flexible transmitter requires two pieces of test equipment: an RF signal generator to create the carrier tone, and an arbitrary waveform generator to create the baseband signal (see Fig. 10.1). The two are then combined using one of the upconversion architectures described in the previous chapter. The mixer can be internal to the RF signal generator or a simple external component. Two representative instruments are described in Table 10.1. Note that the very wide carrier frequency range is not covered instantaneously—switching from one frequency to another takes a few milliseconds.

Arbitrary waveform generators (AWG) for bandwidth below 200 MHz are available from many vendors and can be built into the RF signal generator. This bandwidth is sufficient for the vast majority of SDR applications. Options for very wideband operation are more limited, and two examples are described in Table 10.2. Note that these wideband models are designed to play back a fixed length sequence and do not provide a way to supply an external, contiguous data stream (low bandwidth models do have this capability). However, the existence of these AWGs means that an equivalent DAC can be integrated into an SDR.[2] The wideband AWG can be used to generate the signal directly at the carrier frequency (for $F_c < 10$ GHz), without using an RF signal generator.

---

[1] Desirements are an apt term for parameters that are highly desirable, but not required.

[2] Theoretically, it is possible to take apart the AWG and interface directly to the high-speed DACs.

Table 10.1 Representative RF signal generators

Instrument	Carrier Frequency	Internal Mixer
Agilent E8267D	250 kHz–44 GHz	2 GHz
Rohde Schwartz SMF100A	100 kHz–44 GHz	None

Table 10.2 Representative baseband signal generators

Instrument	Sample Rate	Playback depth	Dynamic Range
Tektronix AWG7122C	24 GHz	64 M	50 dB (10b)
Agilent M8190A	12 GHz	2 G	70 dB (12b)

A power amplifier may be required to complete an SDR transmitter based on standard test equipment. The RF signal generators typically output no more than +20 dBm. This output level is sufficient for emulating handheld devices, but is inadequate for long-distance communications. A very wideband power amplifier would be required to cover the full range of carrier frequencies supported by the signal generator. To the best of the author's knowledge, no such power amplifier is currently available. A set of separate amplifiers can be used instead, with a switch to select the desired band.

10.1.2 Receiver

A very flexible receiver can be constructed using either an oscilloscope or a spectrum analyzer (see Fig. 10.2). The oscilloscope setup closely follows the RFFE receiver architectures described in the previous chapter. The oscilloscope can be used to directly sample the RF signal if the sample rate and dynamic range requirements are met. Alternatively, a signal generator (see previous section) is used to generate the desired carrier frequency and downconvert the signal of interest before it is sampled with an oscilloscope.

Modern oscilloscopes offer very high sample rates, but relatively low dynamic range. Two representative examples are described in Table 10.3. The first oscilloscope belongs to the class of general purpose devices and offers relatively limited bandwidth (well below half of the sample rate). The second belongs to the class of high-speed serial link analyzers and offers a very wide bandwidth. At first glance, it may appear that wide bandwidth oscilloscopes can be used to implement the ideal SDR by sampling directly at RF. However, the limited dynamic range means that adjacent channel interferers will drown out the signal of interest. These devices do indicate the maximum signal bandwidth that can be supported in a modern SDR. Multiple gigahertz of spectrum is allocated for data communications at high carrier frequencies (above 40 GHz). It is possible to construct a RFFE for

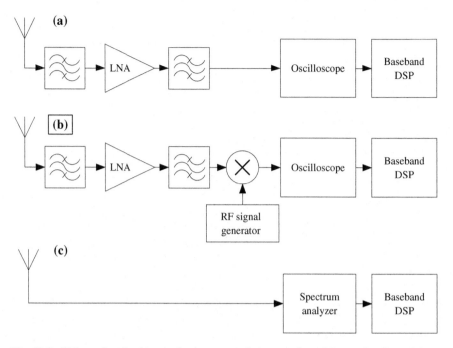

Fig. 10.2 SDR receiver implemented using test equipment: **a** direct-RF sampling, **b** traditional downconversion architecture with IF or IQ sampling, **c** using a spectrum analyzer

Table 10.3 Representative oscilloscopes

Instrument	Bandwidth	Record depth	Dynamic Range
Tektronix DPO7354C	3.5 GHz (40 Gsps)	500 M	< 8 bits
Agilent Infiniium 90000 X	32 GHz (80 Gsps)		< 8 bits

an SDR capable of processing a single signal with >5 GHz of bandwidth[3] using the test equipment described in this and the previous sections.

Oscilloscopes can be used to receive wideband signals, but require appropriate filtering to remove adjacent channel interferers. A spectrum analyzer (SA) combines a signal generator, downconverter, filtering, and a digitizer into a single device [153]. Alternatively, an SA can be thought of as a tunable RF filter (although it is not implemented that way). Modern SA dowconverterts a block of spectrum and then digitizes it. The signal bandwidth is usually below 200 MHz, which is sufficient for the vast majority of SDR implementations (Table 10.4). Some SA output the downconverted analog signal at a low IF frequency and the

---

[3] At the time of writing such signals are only of interest to military users and the nascent ultra-wideband (UWB) standard.

Table 10.4 Representative spectrum analyzers

Instrument	Carrier Frequency	Signal Bandwidth	Dynamic Range
Agilent N9030A PXA	3 Hz to 50 GHz	140 MHz	< 14 bits
Rohde-Schwartz FSQ	20 Hz to 40 GHz	120 MHz	< 14 bits

digitized signal over a parallel digital interface (see Fig. 10.2c). These output interfaces are ideal for integration with the DSP subsystem to create a complete SDR.

10.1.3 Practical Considerations

The test equipment described in this section demonstrates state-of-the-art performance for RF carrier generation, baseband signal generation, and digitization. It is interesting to consider whether a complete SDR can be developed using this equipment and a separate (e.g. FPGA-based) DSP subsystem. The answer is 'sometimes.'

A combination of a vector signal generator (VSG) and a spectrum analyzer with appropriate digital baseband interfaces can be used to implement an SDR for signal bandwidth below 200 MHz and carrier frequencies over a very wide range (Hz to GHz). For example, the Tektronix RSA6000 SA outputs the digitized IQ data over a 16-bit connector. Likewise, an Agilent E8267D can receive digital IQ data for upconversion over a similar interface. The DSP subsystem is then responsible for generating/processing digital samples in real time.

Higher bandwidth signals cannot be processed in real time using available test equipment. The wideband AWGs and oscilloscopes are designed to work on relatively short bursts of data. A pseudo-real-time wideband system with a *low-duty cycle* can be implemented. The transmitted samples are computed in the DSP subsystem and transferred to the AWG (Fig. 10.3). The AWG is then triggered (e.g., based on absolute time) to transmit the burst of data. Likewise, an oscilloscope can be triggered when the received signal level exceeds some threshold or at a specific time (Fig. 10.4). The digitized data is then transferred from the oscilloscope and processed in the DSP subsystem. Oscilloscopes do not provide a way to accurately time-tag the captured samples, but an additional device can be used to record the time of each trigger. Clearly this solution is not suitable for a full-duplex TDMA links since the latency between transmission and reception is quite high (of the order of seconds).

It is *theoretically* possible to disassemble the wideband AWG and oscilloscope and connect the internal DAC/ADC assembly directly to a DSP. Developing a DSP subsystem capable of processing a signal with 10+ GHz bandwidth is extremely challenging, but not impossible with today's technology. Multiple FPGAs or ASICs operating in parallel would be required to keep up with real-time processing.

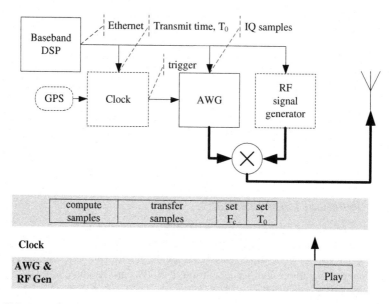

Fig. 10.3 Implementing a low-duty-cycle SDR transmitter using test equipment (dashed blocks are optional)

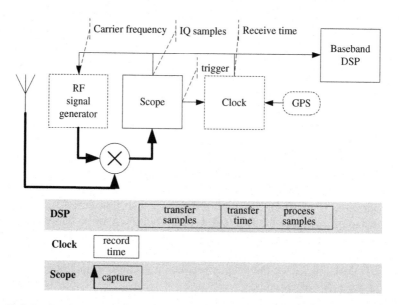

Fig. 10.4 Implementing a low-duty-cycle SDR receiver using test equipment (*dashed blocks* are optional)

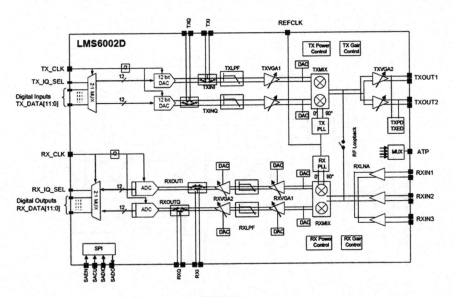

Fig. 10.5 Lime Microsystems flexible RFFE [156]

10.2 SDR Using COTS Components

To make the leap from the laboratory to mass deployment, an SDR must be competitive with single function radios in terms of cost, size, and power consumption. The high-end equipment described in the previous section definitely does not meet these requirements. This section examines components available off-the-shelf in high volumes and at reasonable cost.

10.2.1 Highly Integrated Solutions

Integration of multiple functional blocks on a single microchip is crucial to reducing cost and size of the solution. Some of the earliest work aimed at an integrated SDR RFFE was done at Interuniversity Microelectronics Centre (IMEC). Their latest iteration of a flexible RFFE, shown in Fig. 10.6, supports carrier frequencies from 100 MHz to 3 GHz and signal bandwidth up to 20 MHz [154]. An entire book is dedicated to an earlier version of that chip [155]. Lime Microsystems offers the most flexible commercial RFFE on the market today. This single-chip transceiver supports carrier frequencies from 375 MHz to 4 GHz and signal bandwidth up to 28 MHz [156]. A block diagram of this chip is shown in Fig. 10.5.

Note that both the IMEC and Lime chips use the direct-IF architecture described in Sect. 6.2, and both integrate ADCs and DACs on the same chip. The two chips are more similar than different, but the IMEC chip offers slightly better performance.

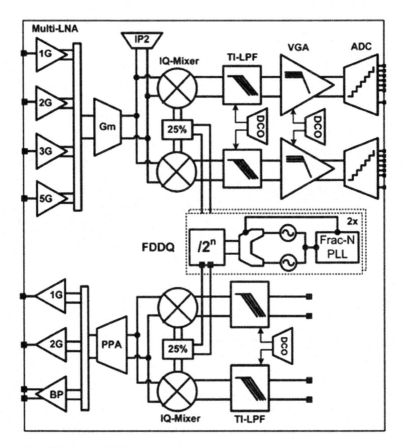

Fig. 10.6 IMEC flexible RFFE [154]

10.2.2 Non-Integrated Solutions

Highly integrated chips are excellent for high-volume, cost-sensitive SDRs, but may not meet requirements for higher end SDRs. Flexible front ends for high-end SDRs can be designed using separate components for each function. The main types of components are:

- Filters and duplexers/diplexers
- Low noise and power amplifiers
- Variable gain amplifiers
- Frequency synthesizers
- Mixers
- ADCs and DACs

The trend to miniaturized surface mount chips has made it very difficult to quickly put together prototypes. A circuit board has to be designed, fabricated, and

Amplifier   Mixer

Fig. 10.7 Connectorized and chip versions of RF components

assembled just to put two chips together.[4] RF components, on the other hand, come in both chip and connectorized versions (Fig. 10.7).

The author has met 'old school' RF designers from large military contractors who were almost unaware of the 'chip' versions, and RF designers from startups developing cell phones who were completely unaware of the connectorized versions. The two worlds seem to exist in parallel. Connectorized devices are prevalent in very high frequency designs (>8 GHz) and are invaluable for quick prototyping. The connectorized versions are up to ten times more expensive and usually operate at higher voltages and burn more power. The connectorized devices are targeted at the high-end market and usually offer better performance than bare chips.[5] For example, a connectorized amplifier typically includes appropriate matching circuits.

10.2.3 Analog-to-Digital Converters (ADCs)

The dynamic range and bandwidth of an ADC converter is often the limiting factor in the flexibility of an SDR receiver.[6] ADCs are semiconductor devices, but their performance does not follow Moore's law—i.e., performance does not double every 18 months. According to [35,157], the resolution-bandwidth[7] product doubles every 4 years and power consumption decreases by 50 % every 2 years.

---

[4] Evaluation boards are sometimes available and can be used during the prototyping stage.

[5] There is, of course, nothing magical about the connectorized devices. Inside the box, they still use the chip components. However, the chips are attended by high-quality passives and the board layout is excellent.

[6] The ultimate limit for ADCs is estimated using the Heisenberg uncertainty principle. Using the equation $\Delta E \Delta t > h/2\pi$, where ΔE is the energy of the smallest resolvable signal (½ LSB), and Δt is half the sampling period (T/2), and h is Planck constant. Assuming a 1 V peak-to-peak input signal and a 50 Ω impedance, the Heisenberg limit yields 12 bits at 840 GS/s. This limit is ∼4 orders of magnitude beyond the state-of-the-art which is currently aperture jitter limited [164].

[7] The number of bits quoted on ADC datasheets is an almost meaningless metric. The effective number of bits (ENOB) is a better measure and is typically 2−5 bits less than the 'nominal' resolution. ENOB is driven by a combination of noise and spurs.

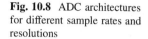

Fig. 10.8 ADC architectures for different sample rates and resolutions

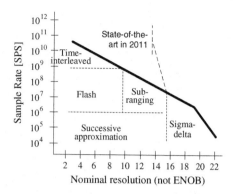

Most modern ADCs are based on one of four architectures that tradeoff resolution and sample rate as shown in Fig. 10.8.[8] An SDR designer does not typically care which architecture is used for the selected ADC,[9] and the discussion below is presented for completeness.

- FLASH. Flash ADCs compare the input voltage to 2^N reference voltages. The closest reference voltage is declared equal to the input. This architecture is very fast, but requires a lot of chip area since 2^N comparators and buffers must be created. Further, it is difficult to accurately generate that many reference voltages (e.g., difficult to match 2^N resistors with high tolerance). This architecture is used for the highest speed (>1 GHz) devices and typically does not provide more than 8 bits of resolution.

- SAR. Successive approximation ADCs compare the input voltage to different generated voltage levels. Starting with a voltage of ½ scale, the ADC generates ever finer voltages in steps of ½ of the remaining scale. A total of N clock cycles are required to achieve N bits of resolution. The approximation stages can be pipelined using N comparators instead of just one. A pipelined SAR architecture increases throughput, but still incurs a latency of N clock cycles. This architecture is used for the vast majority of all ADCs, and covers sample rates up to 100s of MHz and resolution up to 18 bits.

- Sigma-Delta. Sigma-delta ADCs use a single comparator running at a high rate. At each clock cycle, a single bit of data is generated. The bits are then filtered and decimated to create the final output. Filtering allows the ADC to achieve very low noise levels in the band of interest and thus achieve high resolution. These ADCs are used for resolution up to 24 bits, but offer much lower bandwidth. Sample rates are lower because many clock cycles are required for a single measurement. A variation of this architecture is known as delta-sigma.

---

[8] Other architectures such as R/2R have fallen out of favor for commercial ADCs.

[9] In rare instances, the *latency* through the ADC is important. For example, SAR and sigma-delta ADCs are not suitable for systems that require very fast feedback loops and cannot tolerate any delay in the loop.

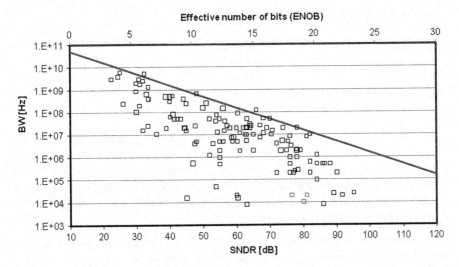

Fig. 10.9 Published ADC performance (1993–2011): bandwidth versus precision (*line* indicates performance limit for sample jitter of 1 ps) [158]

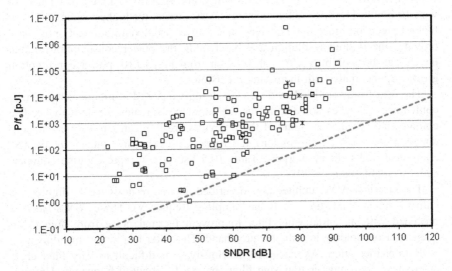

Fig. 10.10 Published ADC performance (1993–2011): power consumption versus precision (*line* indicates 10 pJ/f_s) [158]

- Time Interleaved. Time interleaved ADCs apply the concept of parallel processing by using multiple ADCs simultaneously. The sample time for each constituent ADC is offset relative to others by either delaying the sample clock or the input signal. For example, 10 ADCs sampling at 10 MHz and offset in time by 10 ns are 'equivalent' to a single ADC sampling at 100 MHz. This architecture is only used

Table 10.5 State-of-the-art ADCs in 2011

Device	Sample Rate	SNDR[*]	Nominal bits
National Semi ADC12D1800	3.6 GSps	58	12
Analog AD7760	2.5 MSps	100	24
TI ADS1675	4MSps	92	23

[*] Worst case SNDR is reported over the entire band

for ultra-high sample rates that cannot be achieved with a single device. It is very difficult to accurately time align the ADCs and even more difficult to make sure all ADCs have exactly the same internal gains. Mismatches result in reduced accuracy. For example, time interleaving four ADC with 10 ENOB, may result in an ADC with only 8 ENOB. The high-speed ADCs in modern oscilloscopes are all based on this architecture.

A number of through surveys of ADC have looked at different metrics such as speed-resolution product, power versus speed, speed versus resolution, etc. (see [35] and references therein). An up-to-date compendium of ADC performance data is available in [158]. As can be seen in Fig. 10.9, the current state-of-the-art ADCs can digitize a 1 GHz wide signal with about 9 ENOB and 10 MHz wide signal with 13 ENOB. The near-term goal for high-end ADCs is 10 GHz of bandwidth with 10 ENOB [159]. Note that very few ADCs target the highest sample rates or highest precision, since these devices have limited use. Even if an ADC exists that provides the desired resolution and bandwidth, the power consumption of that device can be prohibitive. As can be seen from Fig. 10.10, power consumption grows exponentially with increasing precision.

The data from [158] is based on chip designs reported at premier engineering conferences. It takes some time for the latest and greatest devices to become commercially available. Some of these devices are fabricated in exotic semiconductor processes (i.e., not silicon) and remain too expensive outside of the test equipment and military markets. Table 10.5 lists a few representative commercially available parts in 2011.

The classical ADC architectures described above are designed to digitize the entire Nyquist band (up to half the sampling frequency). These ADCs *must* be preceded with an antialiasing filter to prevent thermal noise from higher frequencies from folding into the Nyquist band. This filter is a major impediment to SDR implementation. As discussed previously, it is difficult to vary filter characteristics[10]; and most antialiasing filters have a fixed cutoff frequency. Unfortunately, this means that the ADC has to run at twice the filter frequency even if the signal of interest is narrower.

Interesting ADC architectures have been proposed that eliminate the antialiasing filter and/or allow sampling of bandpass signals. Antialiasing filtering can be

---

[10] But not impossible. For example, the highly integrated RFFEs described in Sect. 1.2.1 include tunable antialiasing filters.

Table 10.6 State-of-the-art DACs in 2011

Device	Sample Rate	SFDR*	Quoted resolution
Euvis MD662H	8 GSps	45	12
Maxim MAX5881	4.3 GSps	60	12
TI DAC5670	2.4 GSps	60	14
Maxim MAX5895	500 MSps	73	16
Analog AD9122	1.2 GSps	72	16

integrated into the digitization process as described in [160]. Removing the antialiasing filter is very desirable for SDR implementations. The sigma-delta architecture can be adapted to digitize a narrowband signal centered at a non-zero carrier frequency as described in [161]. A number of the bandpass sampling ADCs have been reported in the last 10 years. However, they have limited applicability to SDR since the signal center frequency is not easily tunable.

10.2.4 Digital to Analog Converters (DACs)

DACs do not receive nearly as much attention in the research community as ADCs.[11] DACs are arguably easier to implement than ADCs. Interestingly, DAC architectures closely mirror those of ADCs:

- A fully parallel implementation uses either 2^N resistors or current sources. This architecture provides the highest sample rate. However, matching the individual sources becomes very difficult, limiting the precision of the DAC.
- Binary weight. These DACs use N resistors or current sources, with each source weighted according to its position in the N-bit word.
- Sigma-Delta. This architecture is very similar to the ADC sigma-delta.
- Hybrid architectures. Most practical DACs use a combination of the above architectures. For example, the MSBs may be implemented full parallel, and the LSBs using the binary weights.

Table 10.6 lists a few representative commercially available parts in 2011 (Fig. 10.11).

10.3 Exotic SDR Components

The test equipment and COTS components described in the previous two sections are available (at a price) to most SDR developers. This section briefly touches on some new and exotic technologies that have not yet made it out of the lab. Some of

---

[11] An informal search of the IEEEXplore database for the past 10 years returned ~4,000 papers on ADCs and ~1,000 papers on DACs.

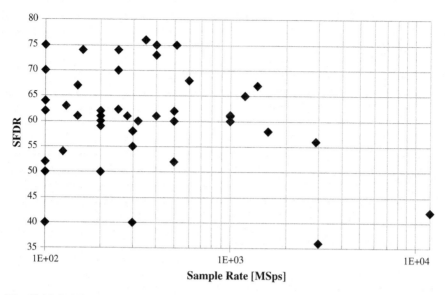

Fig. 10.11 Published DAC performance specifications (1995–2010) [162]

the most interesting research on very wideband data conversion involves the use of photonic components [163]. A comprehensive overview of ADC technology in general and photonic ADCs in particular is provided in [164]. In a generic photonic ADC, the electrical signal that must be digitized is first modulated onto a laser beam. The laser beam is then processed in the optical domain using nonlinear optical components. The processed optical signal(s) are then converted back to the electrical signals using photodetectors. The electrical signals are finally processed using standard electronic ADCs/DACs. The resultant system is not yet practical for commercial SDRs. Optical components are currently quite large (but getting smaller [165]) and expensive. The power consumption of the laser source can be quite high. Photonic ADCs currently provide very wide bandwidth (∼80 GHz) but limited resolution (<4 bits).

Photonic components are also being applied to the RFFE chain [166]. Photonic circulators [167] and tunable filters (see Sect. 1.4) have been reported in the open literature. These exciting developments are mainly of interest to the academic and military communities but will most likely trickle down to the commercial SDR developers in the coming years.

Cryogenics has been used in the design of high-end radios for many years. Improvements in cryo-cooler design are making these components more accessible. Cryo-ADCs offer both lower power consumption[12] and higher resolution bandwidth products than room temperature ones [168]. At least one company is aggressively marketing a cryo-ADC [169] based on a sigma-delta architecture.

---

[12] Neglecting the power consumption of the cryogenic subsystem.

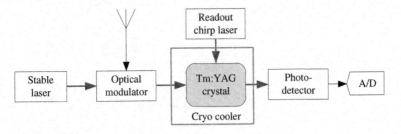

Fig. 10.12 Ultra-wideband photonic/cryogenic spectrum analyzer

According to the marketing materials from HYPRESS, their digital-RF receiver is very close to an ideal SDR. However, a tunable channel select filter is still required for the RFFE.

One particularly interesting system combines photonics and cryogenics. Measuring the power spectrum density (PSD) over a wide bandwidth and doing it quickly is very important for cognitive radio (see Sect. 3.4). The traditional approach uses a wideband ADC to digitize the samples, and then a high-speed FFT engine on a GPP or FPGA to compute the PSD. A few GHz of spectrum can be analyzed using state-of-the-art ADCs. A photonic spectrum analyzer relies on a unique property of a Thulium-doped YAG crystal cooled down to 5°K [170, 171]. This crystal can 'record' PSD of a microwave signal with well over 10 GHz of bandwidth (extendable to 200 GHz using a different crystal [172]). A stable laser is modulated with the received microwave signal. The modulated laser passes through the crystal and 'burns in' the PSD. The recorded PSD is then read out with a narrowband chirped laser and digitized at a rate of up to 1,000 Hz. Note that the ADC need not be particularly fast—100 MHz is sufficient. Thus, a complete 10+ GHz spectrum can be measured every ms with about 60 dB of dynamic range. This ultra-exotic and very expensive ($200 k +) technology is aimed squarely at the military market (Fig. 10.12). As is often the case, the answer to a question "what is possible in today's technology?" is that it depends on how much money is available.

10.4 Tunable Filters

Lack of a tunable RF bandpass filter is the single largest impediment to realization of an ideal SDR receiver. Tunable filters are also desirable, but are not required for SDR transmitters. This section briefly covers existing and emerging technologies for tunable RF filters. A key metric for a filter is its Q factor—the ratio of the bandwidth to the center frequency (Fig. 10.13). The higher the Q, the narrower and "sharper" the filter. For example, to extract a 1 MHz signal at a carrier frequency

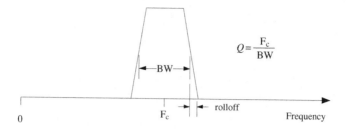

$$Q = \frac{F_c}{BW}$$

Fig. 10.13 Filter Q factor

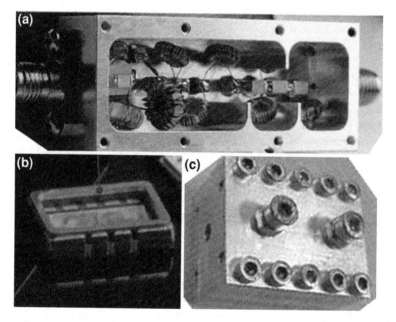

Fig. 10.14 Sample filters **a** RLC, **b** SAW, **c** waveguide

of 1 GHz, the Q should be approximately 1,000. It turns out that designing filters with large Q is difficult. Both the center frequency and the bandwidth of a tunable filter should be continuously and independently variable. An excellent review of tunable filters is available [173].

The main technologies for manufacturing RF filters are:

- RLC (Fig. 10.14a). RLC filters are designed using standard resistors (R), capacitors (C), and inductors (L). These devices can be discrete parts mounted on a substrate (for lower F_c) or traces on a circuit board (for higher F_c). It is difficult to get high Q-factors with RLC filters because many stages of filtering are required. RLC values in each stage must be closely matched to achieve the

Fig. 10.15 Tunable filter
using a filter bank

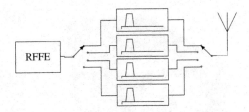

desired frequency response. RLC filters can be tuned by changing the value of one or more components (usually the capacitor).

- SAW (Fig. 10.14b). Surface acoustic wave filters are mechanical (not electronic) devices that happen to resonate at some frequencies and dampen others. SAW filters have very high Q-factors and very sharp rolloff. Unfortunately, they do not work well above 3 GHz and are not tunable.
- Waveguide (Fig. 10.14c). Waveguide (and cavity) filters rely on constructive and destructive interference of EM waves in the filter cavities to achieve the desired response. These filters are typically machined out of single block of metal. The filter size is inversely proportional to the center frequency. Waveguide filters are mostly used for frequencies above 2 GHz since they are very large for lower frequencies. Waveguide filters can be tuned by changing the dimensions of the cavities (e.g., by tightening or loosening strategically placed screws).

Selecting one from a bank of available filters is currently the most popular technique to create a tunable filter (Fig. 10.15). The constituent filters can use any technology. The disadvantages of this approach are clear:

- Multiple filters take up a lot of space.
- Each level of switching causes loss of signal power (less than 0.5 dB per stage), reducing receiver SNR or increasing transmitter power requirements.
- The switches are not ideal, and large interferers can leak into the signal of interest through the 'off' switches.
- The filter is only minimally tunable—bandwidth cannot be changed, and only a few center frequencies are available.

Mechanically tunable filters have been available for decades. In fact, some high-end analog AM/FM radio receivers used tunable filters. Turning the tuning dial actually moved capacitor plates in and out, changing the center frequency of an RLC filter.[13] The center frequency of these filters can be varied contiguously over the filter range (typically one octave). However, the Q factor is fixed and the filter bandwidth increases at higher center frequencies. These filters are valuable for laboratory use but are of limited practical use for SDR development (Fig. 10.16).

---

[13] Most radios just used that variable capacitor to change the output frequency of the synthesizer.

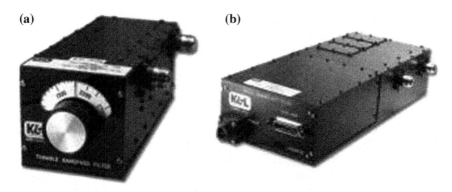

Fig. 10.16 Mechanically tunable filters: **a** manual control, **b** electronic control [174]

Tunable RLC filters are created by using devices with variable capacitance. Two main technologies for electrically (vs. mechanically) tunable capacitance are varactors and ferroelectric films. The dielectric constant of a ferroelectric compound BST (barium strontium titanate) changes with applied voltage. BST can therefore be used to make variable capacitors [175]. At least one company is attempting to commercialize BST technology [176].[14] BST-based tunable filters have been reported for a wide range of center frequencies, from baseband to 30 GHz [177, 178] with high Q-factors. Unfortunately, BST filters require large voltages to tune—up to 100 V. To the best of the author's knowledge, BST filters are not commercially available. Varactors are simply diodes fabricated in standard semiconductor processes. They are inexpensive and easily integrated on the same die as other RF and mixed-signal circuits. Varactor-based tunable filters are used in the single-chip RFFE described in Sect. 1.2.1. Unfortunately, these filters cannot achieve high Q-factors and are not readily usable for carrier frequencies above 500 MHz. A representative example of a varactor-tuned filter is marketed by Hittite [179].

Yttrium-Iron-Garnet (YIG) is a ferromagnetic material that has been used to make tunable filters since the 1960s. A magnetic field is used to change filter properties. Unlike the mechanically tunable filters described above, the bandwidth of a YIG filter is typically fixed (and typically <100 MHz) and does not change with the center frequency. YIG filters have been reported that cover center frequencies from 0.5 to 40 GHz [180]. The passband of these filters is not very flat and requires equalization in the DSP subsystem. YIG filters consume a lot of power to generate the required magnetic field. They are also quite large (typically about 10 cm^3), and therefore not suitable for use in handheld devices. They cannot handle power above 10 dBm, and are therefore not suitable for filtering power amplifier output in a transmitter. Despite their drawbacks, YIG filters are currently the only solution for truly wideband and high-Q filters. A number of vendors are

---

[14] The filters marketed by AgileRF are targeted for baseband antialiasing rather than for RF channel selection.

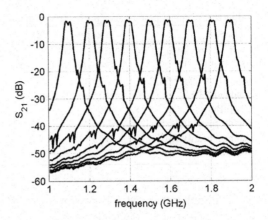

Fig. 10.17 Representative YIG filter tuned over 1–2 GHz range

actively marketing YIG filters [181,182,183], targeting the high-end communications and military markets. Performance of a representative filter is shown in Fig. 10.17. Note the very high Q and a significant ripple in the passband.

The most promising technology for high-Q wideband tunable RF filters is micro-electro-mechanical-systems (MEMS). MEMS-based RF filters use a large number of miniature on/off switches to add and remove capacitors from the circuit [184]. The resistance of each switch is very small, making the overall insertion loss small. A MEMS filter can be thought of as a highly miniaturized version of the filter in Fig. 10.15. Most MEMS filters are not truly continuously tunable, since the number of bandwidth and center frequencies is determined by the number of switches. However, if the tuning resolution (inversely proportional to $\log_2$(#switches)) is reasonably high (>5 bits), the filter can be considered continuously tunable for all practical purposes. MEMS filters can be integrated into a standard semiconductor process, making them potentially low cost, very compact, and low power. Design of these filters is currently a very active research area with over 1,000 papers published in the last 10 years. However, to the best of the author's knowledge, they are not yet commercially available. Transistor-based switches can be used instead of MEMS switches, but offer lower performance.

The technologies described in this section are summarized in Table 10.7.

RF photonics filters rely on nonlinear properties of optical components. These filters first modulate the RF signal on an optical carrier (i.e., laser) and process the laser beam in the optical domain before converting back to electrical. Just like the photonic ADCs described earlier, photonic RF filters have not yet made it out of the laboratory. However, some very exciting development is taking place with *chip-scale photonics* [185,186]. Photonic components are being integrated on a semiconductor chip, making the entire filter relatively compact and low cost. The filters reported so far offer very a wide range of center frequencies, but do not allow narrow bandwidth. The most impressive photonic filter, reported in [187], offers a center frequency of 1–40 GHz and a bandwidth of 250–1,000 MHz. These results are significantly better than prior art and should be treated as 'what will be

Table 10.7 Typical performance parameters of microwave tunable bandpass filters [184]

Parameter	Mech.	YIG	RLC	BST	MEMS
Insertion Loss [dB]	0.5–2.5	3–8	3–10	3–5	3–8
Unloaded Q	> 1000	> 500	< 50	< 100	100–500
Power Handling [W]	500	2	0.2	–	2
Bandwidth [%]	0.3–3	0.2–3	> 4	> 4	1–10
IIP$_3$ [dBm]	High	< 30	< 30	< 30	> 50
Tuning speed [GHz/ms]	Low	0.5–2	1000	–	100
Tuning linearity [MHz]	±15	±10	±35	–	–
Miniaturizable	No	No	Yes	Yes	Yes

possible' rather than 'what is possible.' At this time photonic filters are only used for military (radar) applications, but are definitely a technology to watch as integrated photonics matures.

10.5 Flexible Antennas

Many excellent books are dedicated to antenna design [188]. This section touches on just a few aspects of antenna design as it pertains to SDR. At the most basic level, an antenna is a passive device designed to capture as much *desired* electromagnetic energy as possible. The single largest constraint on antenna design is size. The larger the antenna, the more energy it can capture (as evidenced by the 70 m dish antennas used for deep space communications). Key antenna parameters are:

- Gain. Gain is measured relative to an omnidirectional antennal. A higher antenna gain is achieved by reducing the angle over which energy is received or transmitted. The angle is defined by the antenna beam pattern.
- Frequency response. Antennas inherently act as filters. The antenna efficiency varies with the carrier frequency of the signal. Efficiency strongly depends on the antenna size relative to the wavelength of the signal. Thus, larger antennas are required for lower frequencies and smaller antennas can be used at higher frequencies.

There is a fundamental tradeoff between the width of the antenna beam and the antenna gain. If the user wants to transmit in all directions simultaneously, the gain is low. On the other hand, if transmission is going to a single point, the gain can be high. The parabolic (dish) antennas have a relatively narrow beam and high gain (a function of antenna size), but must be pointed directly at the communications partner. The pointing direction can be fixed or can track the partner. A set of identical non-directional antennas can be used to emulate a directional antenna. The concept of dynamically creating the desired antenna pattern falls in the category of 'smart antennas.' Smart antennas cover a wide range of techniques from

Fig. 10.18 Application of smart antenna technique (beamforming)

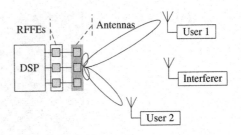

MIMO (see Sect. A.6) to beamforming. Beamforming is a technique that combines appropriately scaled and delayed signals to (from) individual antennas to create the desired beam shape. A beam can be shaped to send all energy to the partner(s), or to avoid (null) an interferer, or achieve a combination of these goals (Fig. 10.18). SDR is an excellent platform for implementing smart antenna techniques. The algorithms are implemented entirely in the DSP subsystem and will not be discussed here.

The relationship between antenna dimensions and its frequency response is one of the biggest hurdles in implementing a wideband SDR. Physics tells us that a large antenna is required to receive low frequency signals. For example, if an SDR needs to process carrier frequencies down to 100 MHz, the antenna size must be of the order of 1 meter. Smaller antennas will work, but result in reduced SNR. There are two approaches to supporting a wide range of frequencies for an SDR antenna:

1. Use a wideband antenna that provides reasonable performance over the entire band.
2. Use a tunable antenna that provides good performance over a narrow band, but allows the band to be tuned over a wide range.

Neither approach is better than the other in all cases. For example, a cognitive radio that must monitor spectrum looking for available bandwidth needs to have a wideband antenna. On the other hand, once a band is selected, a tunable antenna simplifies filtering requirements. A tunable narrowband antenna typically provides higher gain than a similarly sized wideband antenna.

The basic idea behind most tunable[15] antennas is to selectively add and remove segments to achieve the desired geometry, or add capacitors and inductors to achieve the desired resonance frequency. Segments can be added by closing switches between them. The switches can be made using either MEMS [189,190] or semiconductor devices [191] (Fig. 10.19). Tunable antennas are not yet commercially available. A mechanically tunable antenna that uses micropistons instead of MEMS switches was developed by Synotics and is reported in [192].

An entirely different type of antenna has recently been resurrected from obscurity. A plasma antenna uses highly ionized air instead of metal to interact with electromagnetic waves. The size of the ionized volume can be varied by

---

[15] Matching the antenna impedance to the RFFE circuits is a separate task.

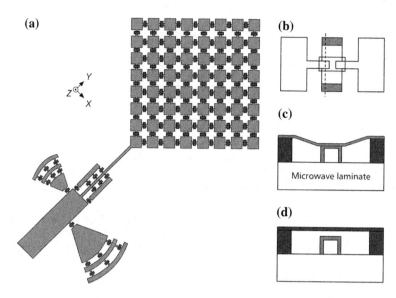

Fig. 10.19 Using MEMS switches to configure antenna geometry [189]: **a** elements and switches, **b** top view of a switch, **c** side view of a closed switch, **d** side view of an open switch

Fig. 10.20 Representative wideband (2–8 GHz) antenna

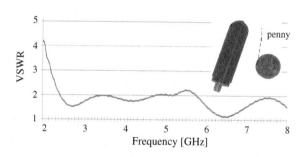

changing the applied voltage (or magnetic field). Plasma antennas are therefore inherently tunable. A comprehensive review is provided in [193] and [194]. Despite their somewhat exotic origins,[16] plasma antennas can be implemented in silicon for low cost and size. At least one company is actively pursuing commercialization of plasma antennas [195].

Since tunable antennas are not yet commercially available, a wideband antenna is the next best thing for an SDR. Fortunately, many wideband antennas are available—from compact designs suitable for handheld devices to large YAGIs. A representative wideband antenna and its VSWR are shown in Fig. 10.20. Note

---

[16] Plasma antennas are of significant interest to the military because they 'disappear' and become invisible when plasma generation is de-energized [272].

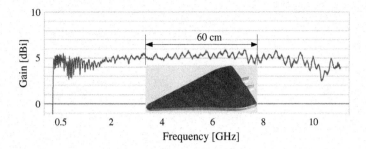

Fig. 10.21 Very wideband antenna (log-periodic) [196]

that VSWR increases (efficiency decreases) at lower frequencies because the antenna is small (just under ½ wavelength at 2 GHz). Another relatively compact antenna that covers 0.4–18 GHz is described in [196] and shown in Fig. 10.21. Large antennas that cover frequencies from 100 MHz to 40 GHz are available. Interesting spiral antennas developed specifically for SDR are reported in [197].

Chapter 11
Development Tools and Flows

Tools used to develop SDR are equally applicable to single-function radios and other complex systems with extensive DSP requirements. The required toolset and development flow strongly depends on the hardware architecture. Unique tools are required for FPGA, ASIC, or DSP implementations. Few decisions engender as much passion as the choice of tools and languages for a new development. Many developers feel almost religious attachment toward their language of choice (C++ vs. Java, Verilog vs. VHDL). In the author's opinion, the choice of a tool within the same class (e.g. C++ vs. Java) makes only a small difference, and the expertise of the key team members counts for a lot. However, switching between classes of tools (e.g. VHDL vs. MATLAB/Simulink) can make a dramatic difference (though not always the expected one).

A highly simplified flowgraph for a typical SDR development process is shown in Fig. 11.1. This flowgraph does not capture any of the optimization/iteration loops that are invariably required in the course of development.[1]

The requirements defined in the first step are used to drive architecture selection. Different SDR architectures (e.g. FPGA, GPP, etc.) are discussed in Chap. 5. The choice of architecture determines which branch(es) in Fig. 11.1 will be taken.

11.1 Requirements Capture

Requirements definition is without a doubt the most important step in SDR development. As discussed in Sect. 4.2, constraining the capabilities of a SDR is nontrivial. Many SDR requirements mirror those for single function radios but with lists provided for each option (e.g. "radio shall support modulations: QPSK, 16QAM; data rates: 0.5–10 Mbps", etc.). Unfortunately, these types of requirements often lead to poorly designed SDRs. The final product does in fact support all the options, but may not be flexible enough to support minor variations. For example, to ensure

---

[1] Ideally, the loops only go between adjacent levels. In practice, it is not unreasonable for an intractable problem encountered at the implementation of the ASIC physical layer to affect the requirements, causing a ripple of changes down the flowgraph.

E. Grayver, *Implementing Software Defined Radio*,
DOI: 10.1007/978-1-4419-9332-8_11, © Springer Science+Business Media New York 2013

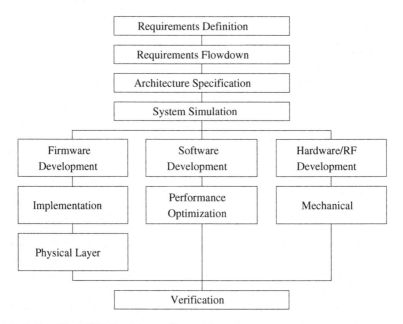

Fig. 11.1 Simplified SDR development flowgraph

flexibility, the modulation requirement is better stated as: "radio shall support an arbitrary constellation described by 2, 4, 8, or 16 points in the complex plane." The more general requirement ensures that the radio can support a 12/4 APSK modulation, which is slightly different from 16-QAM.

In general, many SDR requirements should read "shall *be capable of supporting X...*" rather than "shall support X..." For the first clause, the developer must estimate required hardware to support X. For the latter clause, the developer must actually implement X and it is unreasonable to implement *all* possible waveforms an SDR can support. Of course, these requirements are more difficult to verify. A well-planned SDR development should include a minimum set of waveforms and features at launch, and anticipate new waveforms to be added in the future. Thus, the developer can select an architecture that can readily accommodate the planned features, without requiring that those features be implemented immediately.

The amount of formality and paperwork in the requirements definition phase varies dramatically. A SDR intended for laboratory R&D has few true requirements and a large number of desirements. On the other hand, a SDR meant for military field use must comply with a draconian requirements capture process.[2]

---

[2] A nice set of "requirements for requirements" is articulated in [273]:

Correct (technically and legally possible); *Complete* (express a whole idea or statement); *Clear* (unambiguous and not confusing); *Consistent* (not in conflict with other requirements); *Verifiable* (it can be determined that the system meets the requirement); *Traceable* (uniquely identified and tracked); *Feasible* (can be accomplished within cost and schedule); *Modular* (can be changed without excessive impact); *Design-independent* (do not pose specific solutions on design).

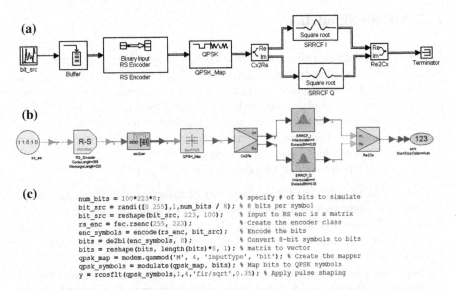

```
(c)    num_bits = 100*223*8;                        % specify # of bits to simulate
       bit_src = randi([0 255],1,num_bits / 8);     % 8 bits per symbol
       bit_src = reshape(bit_src, 223, 100);        % input to RS enc is a matrix
       rs_enc = fec.rsenc(255, 223);                % Create the encoder class
       enc_symbols = encode(rs_enc, bit_src);       % Encode the bits
       bits = de2bi(enc_symbols, 8);                % Convert 8-bit symbols to bits
       bits = reshape(bits, length(bits)*8, 1);     % matrix to vector
       qpsk_map = modem.qammod('M', 4, 'InputType', 'bit'); % Create the mapper
       qpsk_symbols = modulate(qpsk_map, bits);     % Map bits to QPSK symbols
       y = rcosflt(qpsk_symbols,1,4,'fir/sqrt',0.35); % Apply pulse shaping
```

Fig. 11.2 Representing a QPSK modulator with pulse shaping. **a** Simulink. **b** Systemvue. **c** MATLAB

A dedicated software tool is recommended for all but the most relaxed requirements capture processes (a word processor is not an appropriate tool). These tools range from the high-end, such as IBM Rational DOORS [198], to the open source OSRMT [199]. Many of these tools support a graphical language known as the Unified Modeling Language (UML) [200]. UML provides graphical notation techniques to describe how the system should operate, which can then be translated into actionable requirements. For example, UML can be used to describe different *use cases, event sequence diagrams,* or relationships between blocks in the system.

The requirements flowdown process takes the high-level requirements and breaks them down into actionable sub-requirements. For example, a requirement "the transmitter shall support symbol rates up to 100 MSps" can be broken into the following sub-requirements, and each sub-requirement must be traceable to the parent requirement.

1. The DAC shall support sample rate of at least 200 MSps
2. The power amplifier shall have a bandwidth of at least 150 MHz
3. The DSP subsystem shall support a contiguous sample rate of at least 200 MSps.

Each sub-requirement should be linked to a software or firmware file that implements the code to satisfy the requirement. High quality requirements capture tools allow dynamic linking between the source code and the requirement.

11.2 System Simulation

System level simulations allow rapid development of models or prototypes for the entire system or selected critical components. This allows developers to plan and test the big-picture architectures, features, and design parameters before focusing prematurely on final implementation details. System level simulations should be used to verify the performance of selected algorithms under a wide range of expected conditions. For example, signal acquisition would be tested for worst-case frequency and timing offsets and worst-case SNR.

System simulation tools fall into two categories (Fig. 11.2):

1. Code-based tools require the developer to describe the algorithms as a set of sequential commands in a specific language. A simulation may be written in MATLAB, C++, Python, etc. MATLAB is currently the tool of choice for the majority of DSP engineers because of the vast library of predefined functions (toolboxes).[3]
2. Block-based tools require the developer to draw block diagrams to describe the algorithms. Popular block-based tools used today include Simulink, Signal Processing Workbench (SPW), and SystemVue. Simulink is by far the most widely used. Each tool comes with a library of predefined blocks (corresponding to predefined functions for code-based tools). The richness of the blockset is a key differentiator between tools. Additional blocks can be created by either combining existing blocks (hierarchical design), developing new blocks in a code-based tool (e.g. C++) with appropriate wrappers, or purchasing libraries from third-party vendors.

One fundamental difference between code and block tools is the way dataflow is handled. Code-based tools expect the developer to manage data transfer between functions. Block-based tools implicitly provide buffers and flow control between blocks. The difference is apparent in long simulations. A simulation in a code-based tool cannot allocate enough memory[4] to hold all of the data required for the simulation at once. The developer must therefore create a loop to break up the long simulation into smaller chunks. Preserving the internal state for all the functions between each chunk is messy and error-prone. On the other hand, a block-based

---

[3] Interestingly, some developers targeting ASICs strongly prefer C++ for system simulation. This preference may be due to higher simulation speeds offered by C++ versus MATLAB. An ASIC design must be very thoroughly tested across every possible scenario since it cannot be changed after fabrication. The additional effort required to write simulations in C++ is justified by reduced runtimes.

[4] The author came across one project where the system analysts kept requesting more and more RAM for their computers. At one point, a simulation was taking 16 GB and was running very slowly. A quick look at their code revealed that the entire simulation was processed in one step. Eventually the simulations got too long for even high-end machines and the code had to be rewritten to break up the simulation into smaller chunks. The updated code required less than 1 GB of RAM and ran twice as fast.

simulation can process an arbitrary amount of data because each block implicitly saves its own state. If a block-based simulation processes samples one at a time, the simulation cannot take advantage of modern microprocessors' vector computation units and runs slowly. Thus, modern block-based simulations introduce a concept of 'frames' to process data in small chunks. Simulations execute much faster[5] (but still slower than code-based) but become somewhat more difficult to manage if the underlying algorithms require feedback loops that operate on samples.

Despite slower performance, block-based simulators are becoming more and more popular. Proponents argue that a block diagram is easier to understand than equivalent function calls. The simulator framework takes care of many details, making designers more productive. Perhaps the strongest argument for block-based simulations is the newly emerging option to transition a model into implementation (discussed below).

Simulink is the most general purpose of the commercially available block-based simulators. It is also one of the newest entrants into the marketplace. However, tight integration with the immensely popular MATLAB has made Simulink a clear leader.

Signal processing workstation (SPW) is one of the oldest system simulation tools. It was very popular in the 1990s, and is still used by defense contractors with large libraries of legacy designs. In the past decade, SPW has changed hands from Cadence to CoWare to Synopsys. It remains an extremely powerful tool with better visualization and interaction features than Simulink. SPW was probably the first tool to offer a path from simulation to hardware implementation. The library of blocks offered by SPW is unsurpassed by any competitors (SystemVue is a close second) and includes complete solutions for many modern wireless standards. However, SPW costs over ten times more than Simulink, making it accessible only to deep-pocketed companies. More importantly, momentum is definitely with Simulink.

SystemVue is an interesting tool that is arguably easier to use than Simulink. It is targeted specifically for communications system design and has a rich set of communications-specific blocks. Its main strength is a clean interface between digital baseband and RF components. It is the only tool to include a comprehensive set of distortion parameters for fundamental analog blocks (e.g. differential non-linearity properties of a DAC). SystemVue is also unique in that it offers tight integration with Agilent test equipment (e.g. simulation output can be sent directly to a signal generator and the visualization interface is identical to that of a vector signal analyzer). The same GUIs can therefore be used for the entire development process from system simulation to hardware debugging.

---

[5] The author compared throughput of a simulation of a simple BPSK modem with pulse shaping coded using optimized C++, MATLAB, and Simulink. The relative throughput was 13:6:1. Throughput of frame based vs. sample based Simulink simulation was 3:1.

The three products described above are by no means the only available block-based system simulation tools. Tools like VisSim and LabView have a small share of the system simulation pie. In fact, even the GNURadio/GRC platform described in Sect. 8.1 can be used for simulations.

Firmware and software development starts after system simulations are complete and results satisfy the requirements.

11.3 Firmware Development

For the purposes of this section, 'firmware' refers to code intended for implementation on an FPGA or ASIC.

The standard practice in the past was for a systems engineer to hand off a set of requirements, algorithm documentation, and system models to a hardware engineer. The hardware engineer then translated the (usually) MATLAB code to a hardware description language (HDL, VHDL or Verilog). The hardware engineer was responsible for mapping the algorithm to available hardware resources (logic gates, multipliers, memory, etc.) while meeting throughput and latency requirements. The system model generated necessary stimulus files and output files for each algorithm. The hand-coded HDL was then simulated with the stimulus file and the output was compared to the 'golden' output from the system model. This process was time consuming and error-prone. The hardware engineer did not develop the algorithm, making it difficult for him to debug problems. The system engineer is usually not well versed in the hardware 'costs' of different algorithms and can easily come up with a solution that is very inefficient in terms of hardware.[6] Many time-consuming iterations are frequently required to converge on a functional implementation of a hardware-efficient algorithm (Fig. 11.3). The next two subsections present some recent developments that promise to reduce development time by allowing design at a higher level of abstraction.

11.3.1 Electronic System Level Design

Designs are getting more and more complex, but the chip resources available to designers increase even faster. There is a strong effort in industry to combine the roles of system and hardware engineers. This effort is centered on a concept called electronic system level (ESL[7]) design [201, 202]. The goal is to develop firmware

---

[6] A division operator ($\div$) does not typically warrant a second thought from a communications engineer, but dividers are anathema to hardware designers because they are difficult to implement and require a lot of resources.

[7] Also known as high-level synthesis (HLS).

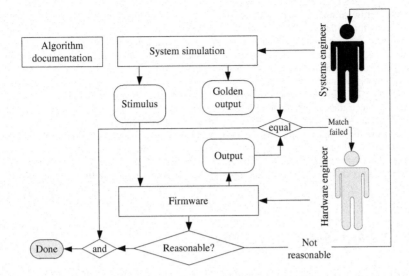

Fig. 11.3 Typical system model → hardware implementation flow

at the system level, bypassing the manual conversion of the system model to HDL. The idea is by no means new—a tool called Behavioral Compiler was marketed by Synopsys as early as 1994 [203] (also see [204] for a comprehensive but somewhat biased history). Unfortunately, the tools were not good enough, and chip resources were too valuable to waste on poorly optimized designs. ESL got a second wind around 2005 with the introduction of a few tools that promised to convert standard (or semi-standard) 'C' code to HDL. Some of the entrants were: Impulse-C, Dime-C, Handel-C, and the only currently surviving option, SystemC. SystemC has yet to be widely adopted and may follow the fate of the others. An extremely promising tool was developed by AccelChip to convert MATLAB code directly to HDL. Unfortunately this tool never gained enough traction, and was discontinued by 2009. It appears that high level synthesis is still "almost there" after 15 years. A recent article [205] highlights the primary difficulty of HLS—the same algorithm can be implemented using very different hardware architectures with easily 1000× difference in design size and performance between them.

> If we are expecting C-to-FPGA to ever behave like a software compiler, we're over-looking an important fact about the difference between hardware and software. For a software compiler, there is always something that could be agreed upon as a "best" solution. Compiler developers can tune away—trying to minimize the size and maximize the speed of the generated code. The right answer is reasonably easy to quantify. Optimization choices made during software compilation have at best a modest effect on the results. For the sake of argument, maybe zero to 20 % plus or minus.[8]

---

[8] The author must disagree with this statement since processor-specific optimizations discussed in Sect. 8.1.1 can provide improvements well over 100 %.

In hardware architecture, however, there is a gigantic range of answers. The fastest solution might take 1000x the amount of hardware to implement as the densest one. The lowest power version might run at a tiny fraction of the maximum speed. The size of the design space one can explore in HLS is enormous. Implementing a simple datapath algorithm in an FPGA, for example, one might choose to use a single hardware multiplier/DSP block for maximum area efficiency—or one might have the datapath use every single available DSP block on the chip—which can now range into the thousands. The cost/performance tradeoff available to the user, then, could be in the range of three orders of magnitude. The "best" answer depends on the user's knowledge of the true design goals, and how those goals map down to the particular piece of hardware being implemented with HLS. Unless the user has a way to express those design goals and constraints and percolate those down into the detailed levels of the design hierarchy, an HLS tool has almost zero chance of guessing the right answer. It is NOT like a software compiler.

11.3.2 Block-Based System Design

Block-based system simulation tools have fared much better in providing a direct path to HDL.[9] SPW has offered an option to convert block diagrams (restricted to using a small subset of its block library) to HDL for over 10 years. More recently, both of the major FPGA vendors have been offering a Simulink blockset that targets their devices (SystemGenerator from Xilinx [206], and DSP Builder from Altera [207]). A Simulink model may use *only* blocks from the blockset to be synthesizable.[10] This clearly limits the usefulness of the simulation environment since the designer cannot leverage the full library of blocks. However, the blocks provided by the FPGA vendors are highly optimized for the target devices. Even more recently, Mathworks has added a feature to Simulink that allows HDL code generation without using vendor-specific blocksets. The HDL Coder [208] generates vendor independent[11] VHDL or Verilog code from a Simulink block diagram using 'standard' blocks. At this time about 150 of the standard Simulink blocks can be used with HDL Coder. It is also possible to use a vendor-specific blockset with HDL Coder, but the design flow is very awkward.

Synopsys Synphony Model Compiler (previously known as Synplify DSP) also provides a path from Simulink block diagram to HDL code [209]. Synphony can generate vendor independent code, and is different from the previously described tools in that it offers powerful architectural transformations of the resultant code.

---

[9] This is hardly surprising since a block-based tool can be 'dumbed down' to mimic a schematics capture tool. Schematics capture was a standard way to develop FPGA designs 20 years ago.

[10] SystemGenerator and HDL Coder allow MATLAB scripts (using a subset of MATLAB syntax and no function calls) to be embedded as synthesizable blocks. This is a very important improvement since it makes putting control logic ('if', 'case', etc.) into a block diagram much easier and cleaner.

[11] Vendor independence is crucial if the design is to be implemented on an ASIC, or to allow the flexibility of moving from one FPGA vendor to another. The downside is that the code is not optimized for a given FPGA architecture.

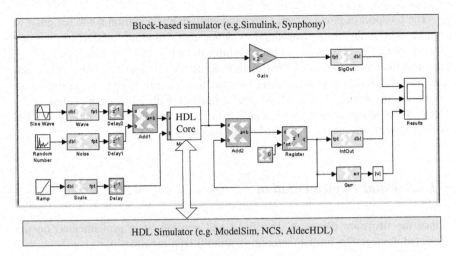

Fig. 11.4 Co-simulation between a block-based tool and an HDL simulator

Transformations allow the same design to be optimized for speed, low power, or logic resources. For example, a filter operating at a sample rate of 1 MSps and 1 GSps implements exactly the same mathematical operation. However, the optimal implementation is clearly very different. The 'holy grail' of ESL is to allow the designer to specify the functionality and let the tools handle the implementation details. This powerful feature comes at a price—Synphony is reportedly at least twenty times more expensive than tools described in the last paragraph. An interesting, but ad-hoc, comparison of three HDL generation tools is provided in [210].

All of the major tools allow the developer to create new blocks. New blocks can be developed in HDL and then linked with the model using co-simulation. Co-simulation requires an HDL simulator to be running alongside the block-based tool, as shown in Fig. 11.4. The two simulators exchange data every clock cycle, significantly increasing simulation time. This approach is absolutely essential for integrating third-party IP cores that are provided in HDL (e.g. a FEC decoder). As the model-to-HDL flow becomes more popular, IP vendors can be expected to provide native simulation blocks with the HDL cores.

Co-simulation can be extended to interface with hardware instead of another simulator. Both Simulink and Synphony support a flow where some of the signal processing is done in the tool and some in an attached FPGA. Simulink supports both JTAG and the much faster Ethernet interface. This hardware-in-the-loop approach can be used to incrementally develop the FPGA firmware. The FPGA can also act as a hardware accelerator, offloading computationally expensive processing from the main GPP (see Sect. 6.2.3.4).

Co-simulation with either an HDL simulator or hardware simplifies unit tests and system-level testbenches, since the test logic does not need to be synthesizable.

The prospect of going directly from model to implementation is too tempting to ignore. The author hears anecdotal evidence that Simulink-to-HDL toolsets are finally being embraced by industry. Unfortunately, with so many completely incompatible players (which also include SystemVue[12] and LabView), choosing one is a major decision. Evaluating multiple tools is a time-consuming endeavor. Choosing the wrong tool can leave the developers with a large but useless code-base. With Simulink a defacto standard in the communications industry, HDL Coder is probably the tool most likely to succeed.[13]

11.3.3 Final Implementation

Once the firmware has been developed and verified, it is implemented on the selected device. The implementation is a multi-step process:

1. The HDL code is synthesised to a set of primitive logic gates (i.e. AND, OR, etc.). The synthesis tool performs some low level optimization (e.g. move registers around to break up long paths, duplicate registers, eliminate redundant logic). FPGA vendors typically provide a synthesiser as part of their tool suite. Both ASIC and FPGA developers can also use third-party synthesisers (e.g. from Synopsys). These dedicated synthesisers can provide slightly better quality of results for FPGA designs than the vendor provided tools (on the order of 10 % reduction in size or improvement in throughput[14]).
2. The logic gate netlist is then mapped to the device fabric (look up tables for FPGAs, standard cells for ASICs). The mapping is tightly coupled to the placement of primitives and routing between them.
3. The placed and routed design is ready for back-end processing. Configuration file generation is the only back-end process required for FPGA designs. ASIC designs require many back-end process steps to ensure that all the design rules have been followed and the fabricated chips have good yield.[15]

---

[12] Hardware implementation features in SystemVue seem to be an afterthought. The HDL blockset includes only a few primitives (e.g. add, subtract, register).

[13] The author's very guarded opinion is that HDL Coder is a reasonable choice for ASIC developers and a *combination* of HDL Coder and a vendor-specific blockset is the right choice for FPGA developers.

[14] Synthesisers can only optimize hand-written HDL code, and cannot improve performance of IP cores provided as netlists. In the author's experience, the critical path (longest delay) in FPGAs usually happens inside the complex vendor-provided cores.

[15] ASIC back-end processing requires highly specialized expertise and any mistake carries an extremely high cost. If a company makes fewer than four tape-outs a year, this step should be outsourced to a turnkey ASIC shop [293].

11.4 Software Development

Readers are most likely already familiar with standard software development flow. This section does not provide a comprehensive review of the software development process, but highlights a couple of aspects. Since 'software' is the first word in the acronym SDR, it is hardly surprising that software is often the most complicated and labor-intensive subsystem.[16]

11.4.1 Real-Time Versus Non-Real-Time Software

Software can be split into two classes:

1. Real-time: Real-time software must process data and events within a tight time constraint. Most of the software for an all-software radio is real-time (e.g., received samples must be processed at least as fast as they come into avoid overrunning input buffers). Many modern wireless standards require tight and accurate timing for transmitting messages (e.g., an uplink packet has to be sent within x ms of receiving a downlink packet).
2. Non-real-time: Non-real-time software does not have tight constraints on responding to events. For example:

   a. User interfaces have to be responsive but there are no hard requirements.
   b. Low data rate standards allow messages to be precomputed and then sent out in a burst.

An operating system (OS) provides a layer of abstraction between the GPP and the application software. An OS is used for all but the simplest applications. Just like the application itself, the OS can be real-time (RTOS)[17] or standard, non-real-time (SOS). A common misconception is that real-time applications require an RTOS. That is not always the case.

Consider an all-software satellite TV recorder. The input data is sampled at, say, 10 MSps and must be processed at that rate to avoid dropping video frames. Received and decoded data is saved to disk. The software is therefore clearly real-time. However, there is no hard constraint on when a particular frame is saved to

---

[16] In the author's experience, most large schedule slips and cost overruns are due to underestimating the software complexity.

[17] A key characteristic of an RTOS is the level of its consistency concerning the amount of time it takes to accept and complete an application's task; the variability is jitter. A *hard real-time* operating system has less jitter than a *soft real-time* operating system. The chief design goal is not high throughput, but rather a guarantee of a soft or hard performance category. A RTOS that can *usually* meet a deadline is a soft real-time OS, but if it can meet a deadline *deterministically* it is a hard real-time OS.

disk. As long as the input buffers are large enough, the software can run on a non-real-time OS.

For example, the GNURadio and OSSIE frameworks discussed in Chap. 8 are designed to run on regular Linux and Windows SOS, but can process real-time data. The price for high process switching jitter in SOS is larger input and output buffers. Since mixed-signal subsystem always runs in real-time, the buffers must be large enough to sustain data flow while the application is not active. If the mixed-signal subsystem supports timestamps, an SOS application can provide very accurate packet transmission times (see Sect. 6.1). An RTOS is required if the application has to respond to received data within a tight time constraint. For example, the UMTS standard requires a handset to update its output power in response to commands from the basestation at a rate of 1500 Hz. The update interval, 0.6 ms, is too short to reliably hit with an SOS. An RTOS is therefore required.

The decision to go with an RTOS or SOS should be made only *after* requirements have been defined. If the latency and jitter requirements are 10 ms or more, an SOS will work fine, but an RTOS is required to achieve μs values. A comparison of different OS options in the context of SDR is presented in [211]. Since RTOSs are designed for embedded applications, they have smaller memory footprints than SOSs. An obvious question is: why not always use an RTOS instead of a SOS?[18] Some of the disadvantages of an RTOS are:

- Slightly reduced performance. The timer interrupts used to ensure hard-real-time constraints add overhead. Keeping uninterruptible critical zones in the kernel small to ensure predictable response times also decreases average throughput.
- Less convenient development. The development tools for RTOS have improved dramatically over the past 10 years. Commercial vendors such as Wind River provide excellent integrated development environments [212]. Despite the improvements, the tools are still not as powerful or user friendly as for SOS. This is usually the strongest argument against using an RTOS.
- Cost. Commercial high-quality RTOS (e.g. VxWorks) can cost well over $10 k. However, open source alternatives do exist.

Hybrid SOS/RTOS solutions are available for developments that require features of a SOS (e.g. a graphical user interface) and an RTOS (e.g. INtime for Windows [213] and Lynuxworks for Linux [214]). These solutions typically run a SOS as a separate process in an RTOS kernel. Alternatively, an RTOS kernel is tied to the lowest level of the SOS kernel. The processes running in the RTOS world can seamlessly communicate with the SOS world.

---

[18] Many old-school engineers also question why any OS is required. Very simple embedded designs can get away without using an OS, but almost any SDR is complex enough to require an OS to manage memory allocation and threading.

11.4.2 Optimization

High power consumption and low throughput are the main reasons radios are not all developed using software on GPPs. As discussed previously, GPPs are just not well suited for digital signal processing. Meeting real-time throughput is always a major challenge for developing all-software radios. At any complexity point, the desire is always to use the smallest, cheapest and lowest power GPP capable of processing the data. GPPs are getting more complex, but the clock rates are not increasing. GPPs now sport multiple cores, single instruction multiple data (SIMD) features, and even dedicated hardware accelerators for encryption. Unfortunately, taking advantage of these features using standard programing language constructs is not easy. Compilers are just not good enough to convert hand-written high level code to efficient SIMD instructions. A program that serially executes all the signal processing cannot take advantage of multiple cores. Software optimization for SDR relies on the following three components[19]:

- Profiling. It is unreasonable and inefficient to optimize every function in a large SDR.[20] A profiler allows developers to pinpoint the 'hot spots' and concentrate on optimizing them.[21] A typical profiler output is shown in Fig. 11.5. In this particular example, the profile shows that the division operation (line 16) takes about two times longer than an addition (line 5).
- Multi-threaded programing. Signal processing tasks must run in multiple threads with as little data exchange between threads as possible.
- Function libraries optimized for the GPP. As mentioned previously, a compiler cannot always generate the most efficient code. Hand-optimized (often written in assembly) libraries of frequently used functions are available from GPP and SPU vendors. The library developers are intimately familiar with the hardware architecture and take the time to optimize each function. For example, an FIR filter function from the Intel library runs about four times faster than one written in C and compiled with an Intel compiler. ATLAS is an open source alternative to vendor specific libraries [215].

Consider the following example: An all-software SDR receiver is performing acquisition, tracking, demodulation and FEC decoding. Functionality is verified using stored data files, but the receiver cannot meet real-time throughput. Profiling identifies that FEC decoding takes 80 % of the time. The decoder algorithm is rewritten using optimized libraries and runs twice as fast. The receiver is now about 1.7 times faster. If the throughput is still too low, FEC decoding is spawned

---

[19] A superb treatise on optimization is available in [291].

[20] "Premature optimization is the root of all evil," D. Knuth [290].

[21] Common sense is not sufficient. The author came across a design where the generation of Gaussian noise took significantly more computational resources than a Viterbi decoder. Another time throughput decreased by a factor of 4 after a minor code change. A profiler identified a deeply buried function that was logging data to disk and taking most of the CPU resources.

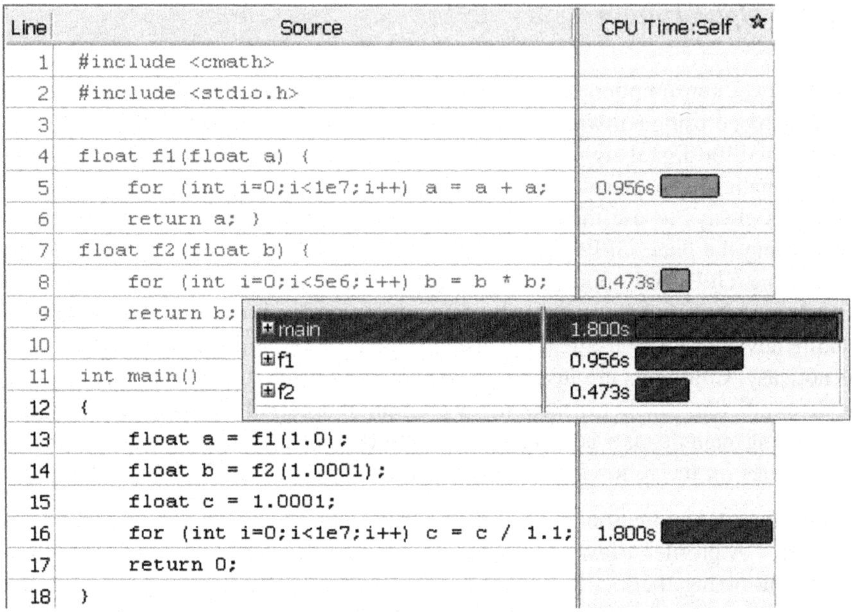

Line	Source	CPU Time:Self ☆
1	`#include <cmath>`	
2	`#include <stdio.h>`	
3		
4	`float f1(float a) {`	
5	`    for (int i=0;i<1e7;i++) a = a + a;`	0.956s
6	`    return a; }`	
7	`float f2(float b) {`	
8	`    for (int i=0;i<5e6;i++) b = b * b;`	0.473s
9	`    return b;`	
10		
11	`int main()`	
12	`{`	
13	`    float a = f1(1.0);`	
14	`    float b = f2(1.0001);`	
15	`    float c = 1.0001;`	
16	`    for (int i=0;i<1e7;i++) c = c / 1.1;`	1.800s
17	`    return 0;`	
18	`}`	

Inset:

⊞ main	1.800s
⊞ f1	0.956s
⊞ f2	0.473s

Fig. 11.5 Output of a profiler showing time used for each line of source code and the time spent in each function (*inset*)

into a separate thread and executes in parallel to the rest of the signal processing. The receiver is now about four times faster. If the throughput is still too low, and the processor has idle cores, more threads can be dedicated to FEC decoding. Otherwise, hardware acceleration can be considered (see Sect. 6.2.3.4).

11.4.3 Automatic Code Generation

Source code for SDR software is still mostly written by hand. However, automatic code generation tools are becoming available. Many of the tools described in Sect. 11.3 can be used to generate either HDL or C code. In fact, since the tools are themselves just software, it is possible to run an entire real-time radio inside a modeling tool.[22] Unfortunately, the tools run relatively slowly compared to optimized software.

Automatic code generation can be used to create the high-level architecture while the low-level functions are implemented by hand. UML allows the developers to describe class relationships and basic execution flow, and is generally

---

[22] Simulink comes with an example design that implements an FM radio receiver capable of running in real-time on a moderately powerful computer.

accepted as the best approach for architecture-level code generation. It is not meant for describing the signal processing inside the classes. As described in Sect. 11.1, UML can also be used for requirements capture. The same diagrams can then be translated into skeleton code. A major advantage of this approach is ease of tracing a requirement to the code that satisfies it. A case for using UML for the development of SDR is presented in [216].

Chapter 12
Conclusion

Any new design of a reasonably sophisticated radio should incorporate certain SDR features. The additional up-front time and costs incurred to add flexibility and programmability will be recouped by reducing the number of design changes later and a potentially larger market.

The main takeaway of this book is: applying more DSP to make better use of the available spectrum, current channel conditions, and the transmit power is almost always a winning proposition. Even non-SDR single-function radios implement many adaptive algorithms. An old receiver adaptively estimates only timing and frequency offset. A modern receiver adapts for channel conditions, RFFE distortions, amplifier nonlinearity, available battery life, etc. Many of these algorithms are implemented in software. The key difference an SDR brings to the table is the ability to change the algorithms themselves, not just the values of the knobs the algorithms control.

As transistors keep getting faster, smaller, lower power, and cheaper, GPPs will become a better platform for SDR development. DSP will consume a smaller portion of the overall radio power budget (still dominated by the transmit power). The flexibility and ease of development will make all-software radios executing on a GPP the dominant SDR architecture.

Modern FEC algorithms combined with smart antenna technology squeeze almost every bit of throughput from the available spectrum (approaching the Shannon limit). A provocatively titled paper "Is the PHY Layer Dead?" was published in 2010 and generated a firestorm of comments [217]. Because many wireless links are operating close to the Shannon limit, there can be little doubt that we are at the point of diminishing returns with every new physical layer algorithm. Does this make SDR any less relevant? Not at all. SDR is the technology that enables developers to take advantage of the latest physical layer algorithms and is therefore just as important even if no new algorithms are developed.

E. Grayver, *Implementing Software Defined Radio*,
DOI: 10.1007/978-1-4419-9332-8_12, © Springer Science+Business Media New York 2013

Appendix A
An Introduction to Communications Theory

Wireless communications is a relatively mature field and has been covered in many excellent books. Classical works by Proakis [1], Sklar [2], Rappaport [218], Haykin [219] and others provide cohesive in-depth coverage of this large and complex topic. The reader is strongly encouraged to become very familiar with one of those tomes before continuing with this book. The field of software defined radio is addressed in the excellent book by Reed [220]. A very accessible introduction to implementing SDR is presented in [221, 222]. A solid understanding of basic calculus and linear algebra are prerequisites to tackling any of these books.

This chapter does not presume to be a substitute for the books cited above. Instead, the author will attempt to give a bird-eye-view of a few aspects of wireless communications. The goal of this chapter is to serve as a refresher to some readers and introduce readers who are new to the field to some of the concepts and terminology used throughout the book. Many concepts discussed below are tightly coupled, making it difficult to introduce them in a sequential order. The reader is encouraged to read over the entire appendix and then return to the sections of interest.

Although wireless communications is a highly mathematical field, the author will attempt to use as few equations as possible. The reader should be able to develop some intuition and then delve deeper into topics of specific interest. Communications is a growth field and you may be coming from a different technical field.[1] After a few months of intensive study it is possible to get comfortable enough with basic communications theory to start implementing SDRs.

A.1 Information

> The fundamental problem of communication is that of reproducing at one point, either exactly or approximately, a message selected at another point.
>
> Claude Shannon 1955

---

[1] My first manager came to communications from applied physics and many of my colleagues majored in fields other than communications.

E. Grayver, *Implementing Software Defined Radio*,
DOI: 10.1007/978-1-4419-9332-8, © Springer Science+Business Media New York 2013

All communication begins with information that must be transmitted from point A to point B.[2] The formal definition of information hinges on the likelihood (probability, $p()$) of each transmitted message. In other words, information is determined by the entropy of the messages. The amount of entropy, H, measured in bits, is defined in the equation below. A message, x, is one of a set of messages,[3] X. A message could be anything from a voltage level to a colored flag.

$$H(x) = -\sum_{x \in X} p(x) \log_2 p(x)$$

Throughout this book we assume that data to be transmitted consists of entirely independent (random) messages. For example, there are two messages in X—one labeled '0' and the other labeled '1'. One of these messages is transmitted with probability of ½. In that case, each message carries exactly 1 bit of information:

$$H(x) = -\left(\frac{1}{2}\log_2\frac{1}{2} + \frac{1}{2}\log_2\frac{1}{2}\right) = 1 \text{ bit}$$

How about ship-to-ship communications that relies on a sailor waving one of four different colored flags?

$$H(x) = -\left(\underbrace{\frac{1}{4}\log_2\frac{1}{4}}_{\text{red flag}} + \underbrace{\frac{1}{4}\log_2\frac{1}{4}}_{\text{green flag}} + \underbrace{\frac{1}{4}\log_2\frac{1}{4}}_{\text{yellow flag}} + \underbrace{\frac{1}{4}\log_2\frac{1}{4}}_{\text{blue flag}}\right) = 2\text{bits}$$

In general, assuming X consists of 2^M equally probable messages, each message carries M bits of information.

$$H(X) = -\sum_{x \in X} p(x) \log_2 p(x) = -\sum_{1}^{2^M} \frac{1}{2^M} \log_2 \frac{1}{2^M} = -\log_2 \frac{1}{2^M} = M\log_2 2 = M$$

A.2 Frequency and Bandwidth

Frequency and bandwidth are two key concepts used to describe any wireless communications system. The electromagnetic *spectrum* is a set of frequencies from zero to infinity. Humans cannot sense electromagnetic waves at most frequencies, making the difference between a 2 GHz and a 4 GHz frequency somewhat abstract. However, our eyes easily distinguish between frequencies in the visible

---

[2] Only point-to-point links will be discussed for simplicity. Most (but not all) results generalize to point-to-multipoint and multipoint-to-point as well.

[3] Only digital communications are addressed in this book. Each message conveys a set of bits. Messages are also known as *symbols*. Thus, the set X is discrete and finite. The set X is contiguous for analog communications.

Fig. A.1 Signal frequency
and bandwidth

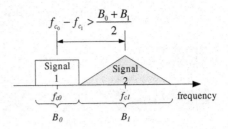

range and we associate an electromagnetic wave at a frequency of ~ 650 THz with the color blue and at a frequency of ~ 450 THz with the color red.

Wireless signals are transmitted at a certain frequency, known as the carrier frequency and denoted, f_c. With rare exceptions a signal occupies not just the carrier frequency, but adjacent frequencies as well. Continuing with the color analogy, a signal is not a single shade of pure red, but covers many shades of red. The range of frequencies occupied by a signal is known as the signal bandwidth and denoted, B. The reason carrier and bandwidth are so important is that two signals cannot easily occupy the same frequency range.

> Consider two sailors standing next to each other and attempting to send different messages to a distant ship. If both sailors are using red lanterns to modulate the data using on-off keying (OOK), the receiver will not be able to tell them apart. However, if one of the lanterns is blue instead of red, the receiver can easily tell which lantern is on at any given time. The human eye can distinguish quite a few colors and many different colored lanterns can be used to simultaneously transmit data.

This example demonstrates why different signals must typically be transmitted on different carrier frequencies.[4]

> Consider a poorly built lantern with glass that ranges in color from blue to green rather than being a single pure color. The lantern now occupies a wider bandwidth. A green lantern cannot be used next to the 'blue-green' one since it will be impossible to tell the two apart. A wider bandwidth lantern reduces the number of other signals that can be transmitted at the same time.

This example demonstrates that the difference between carrier frequencies of two signals must be at least as large as half of the sum of the signals' bandwidths (Fig. A.1).[5]

There are multiple definitions of bandwidth, with almost fanatical adherents to each definition.[6] The most popular are:

---

[4] The reader is no doubt familiar with techniques that allow many signals to share the same bandwidth. Some of these techniques will be discussed later in the book.

[5] If the signals are spaced closer than that, energy from one signal distorts the other. This process is known as adjacent channel interference.

[6] The author has come across wireless systems that are poorly designed because of a lack of agreement on a definition of bandwidth. In one example, a satellite transmitter had to be redesigned because the developers assumed 3 dB bandwidth while the ITU frequency mask was based on 99 % bandwidth.

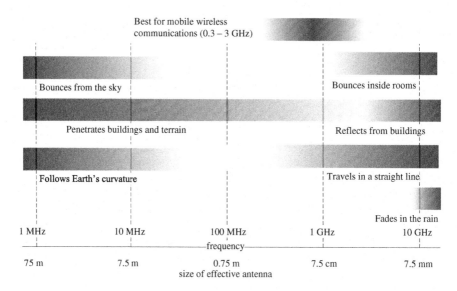

Fig. A.2 Characteristics of different spectrum bands

- [B$^{X\ \%}$] Bandwidth is defined such that X % of the signal power is contained within B Hz. 99 % bandwidth is used most often in regulatory documents.
- [B$^{-X\ dB}$] Bandwidth is defined such that the power level outside B Hz is at least X dB[7] below the peak power level. A special case of this is the 3 dB bandwidth, also known as half-power bandwidth.
- [B$^{X\ BER\ dB}$] Bandwidth is defined such that removing all signal power outside of B Hz does not degrade the AWGN BER performance (see Sect. A.7.1) of the signal by more than X dB.

Although the electromagnetic spectrum is infinite, the subset of frequencies usable for wireless communications is quite constrained. Different frequency ranges are best suited for different communications applications. For example, signals transmitted at lower frequencies can travel longer distances without being attenuated than signals at higher frequencies. Some frequencies can go through walls or other obstructions, while others cannot. Figure A.2 summarizes a few important characteristics of different frequencies.

The 'desirable' frequencies are heavily used and allocation of frequencies to users is strictly managed by the FCC in the United States [16] and equivalent organizations around the world. For example, the best frequencies for mobile broadband (i.e. smartphones) lie between 0.5 and 3 GHz. At these frequencies the waves travel in a straight line, can penetrate into buildings to some extent and do not require very large antennas. The value of the desirable parts of the spectrum can be translated into

---

[7] Communications engineers use logarithmic units—Decibels (dB)—to efficiently describe values that vary over many orders of magnitude.

dollars by looking at the results of recent FCC auctions.[8] The relationship between the amount of spectrum allocated to a user and how much data throughput that user can achieve is discussed in the next section. The goal of much of the SDR technology discussed in this book is to transmit as much data as possible using the least amount of spectrum (bandwidth) while obeying all the other constraints.

A.3 Channel Capacity

The theoretical foundation for communications theory was laid down by Claude Shannon in 1948. Starting with the definition of information, Shannon derived an upper bound on the amount of information that can be reliably[9] transmitted over a communications channel. The general formulation of the result is beyond the scope of this book and is elegantly addressed in chapter 7 of [1]. Only a subset of the theory applicable to a simple AWGN channel (see A.9.1) is considered in this section. The capacity of an AWGN channel is given by

$$C = B \log_2(1 + \text{SNR}) = B \log_2\left(1 + \frac{P}{N}\right) = B \log_2\left(1 + \frac{P}{BN_0}\right)$$

where

B is the bandwidth of the signal
P is the average power of the received signal
N_0 is the noise power density and N is the total noise in the signal bandwidth

This well-known result applies to the simple AWGN channel with a single transmit and a single receive antenna and with the channel gain constant over the bandwidth of the channel. Shannon's result proved that this capacity was theoretically achievable, but provided little insight into how it can be achieved in practice. At the time his work was published, communications links were achieving at best 10 % of the predicted capacity. It took another 40 years for coding theory to develop techniques that now allow us to operate almost at capacity (see Sect. A.8).

Let us consider a few examples to develop some intuition about what this fundamental result really means. We first observe that capacity grows logarithmically with the signal power.

Consider a link with $B = 1$ Hz, $N_0 = 1$ W/Hz, $P = 0.1$ W. The capacity of this link is $C = 1 \log_2\left(1 + \frac{0.1}{1 \times 1}\right) = 0.14$ bps. What happens if we double the transmit power? Capacity increases almost 2-fold to 0.26 bps. What if the transmit power is

---

[8] Over $60 billion has been paid by companies to the US treasury over the past 15 years for the rights to operate in a few spectral bands.
[9] With arbitrarily small error probability of error.

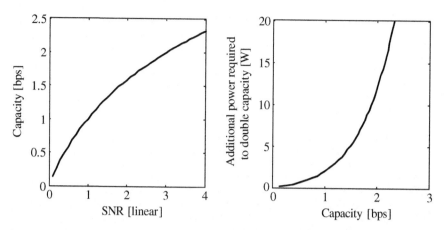

Fig. A.3 AWGN channel capacity and power requirements

1 W? Capacity is then exactly 1 bps. How much extra power would be required to double this throughput? The total power required to achieve 2 bps is 3 W (SNR = 3)—3 times more than before. The point is clear—the amount of power required to increase throughput grows exponentially! It takes a lot of power to increase bandwidth efficiency above 1 bps/Hz (due to the shape of the *log* function) (Fig. A.3).

Let us now consider the effect of available bandwidth, B, on the capacity. At high SNR, capacity increases almost linearly with bandwidth. Of course, as the bandwidth increases, SNR decreases since more noise enters the system.[10]

Consider the same link as in the example above: with $B = 1$ Hz, $N_0 = 1$ W/Hz, $P = 0.1$ W. How much bandwidth is required to double the capacity? Looking at Fig. A.4, we observe that capacity increases slower than logarithmically as bandwidth asymptotically approaches a constant. Thus, capacity cannot be doubled by simply increasing bandwidth without increasing power! This region of the capacity curve is known as *power constrained*.

Now consider a link with $B = 0.1$ Hz, $N_0 = 1$ W/Hz, $P = 1$ W. SNR is 10. When SNR is relatively high, increasing bandwidth helps increase capacity. For example, increasing bandwidth by a factor of 4 to 0.4 Hz approximately doubles the capacity. A corresponding increase in capacity while keeping the bandwidth at 0.1 Hz requires about 15 times more power.

These examples demonstrate an important point about channel capacity. At low SNR (<3 dB), increasing power is an effective way to increase capacity. However, at high SNR (>10 dB), small increase in capacity requires a very large increase in power. Likewise, increasing bandwidth is not effective at increasing capacity at low SNR, but becomes effective at high SNR.

---

[10] Exercise for the reader: analyze the asymptotic behavior of the channel capacity as bandwidth tends to infinity.

Fig. A.4 AWGN capacity
and bandwidth requirements

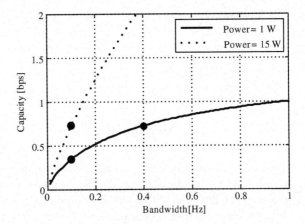

For many wireless links, both power and bandwidth are constrained by applicable standards and regulatory bodies. A relatively new technology, known as MIMO, has been developed over the last two decades to overcome the bandwidth limitation.

A.4 Transmitting Messages

Messages can be transmitted by varying any observable parameter. Dozens of different transmission techniques and message formats have been developed over the years. The conversion between an abstract message and an observable outcome is known as modulation. The observable outcomes are known as symbols. Symbols are transmitted at regular intervals (at a constant rate, known as the symbol rate, R). The bandwidth of a signal is proportional[11] to its symbol rate, $B \propto R$. The proportionality constant depends on the modulation. This section will mention some of the more frequently used modulations.

A.4.1 On-Off Keying

Perhaps the simplest modulation scheme is on-off-keying (OOK). At any symbol interval, a signal is either transmitted or not. Presence of the signal can be mapped to a '1' and absence of a signal can be mapped to a '0.' The signal itself can be anything:

- Smoke rising from a fire ('1') or blocked by an animal hide ('0')
- A flash of light reflected from a mirror ('1') or no flash ('0')

---

[11] For most modulations. Exceptions include PPM described in Sect. A.4.6.

  – Modern optical communications is a logical extension of this ancient approach. Even today, most commercial optical systems rely on simple on/off pulses from a laser.

- A whistle ('1') or silence ('0')
- Nonzero voltage level on a wire ('1') or ground potential ('0')

A.4.2 Frequency Shift Keying

Frequency shift keying (FSK) is another very simple modulation scheme that has been in use for hundreds of years. FSK symbols differ from each other by the frequency of the signal. The oldest example is using different colored lanterns. Different colors of light correspond to different frequencies of the electromagnetic wave (e.g. blue light is ~ 650 THz, while red light is ~ 450 THz). Whistling or playing different musical notes uses different frequencies of sound to transmit different messages. Practical FSK modulation uses 2, 4, 8, or 16 closely spaced frequencies. The spacing between adjacent frequencies relative to the symbol rate is known as the modulation index. The phase of the signal can remain contiguous between successive symbols (continuous phase FSK, CPFSK) or start at an arbitrary or random phase at the beginning of each symbol. Since FSK is relatively easy to generate and detect it is still used for low-data rate links such as satellite command and control.

A.4.3 Continuous Phase Modulation

The continuous phase FSK modulation described above is one of a general class of continuous phase modulations (CPM). Data bits are encoded by smoothly changing the phase of the output signal from the previous symbol to the new symbol. Signal amplitude remains constant. The shape of the phase as it changes from one symbol to the next (phase trajectory) depends on the specifics of the modulation. The contiguous phase trajectory means that the symbol cannot be determined solely from the final phase of the signal. The symbol depends on the final phase and the phase at the beginning of the symbol. The phase at the beginning of the symbol is determined by the cumulative total phase of all previous transmitted symbols. Practical systems design the phase trajectory such that only a finite number of previous symbols have to be considered.

Perhaps the simplest type of CPM is minimum shift keying (MSK). MSK is also a special case of CPFSK, with two frequencies and the modulation index equal to 1/2.

The most frequently used type of CPM is GMSK—modulation adopted for GSM (second generation cellular telephony standard), Bluetooth, and many others. The phase trajectory for GMSK is determined by passing the input data through a Gaussian filter which creates a smooth phase trajectory.

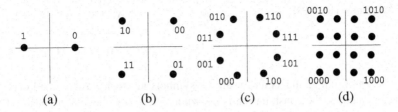

Fig. A.5 Standard APSK constellations: **a** BPSK, **b** QPSK, **c** 8PSK, **d** 16QAM

A.4.4 Amplitude and Phase Shift Keying

The vast majority of modern modulation schemes rely on amplitude and/or phase shift keying. These modulation schemes do not have obvious manifestations like smoke signals or colored flags. Instead, they should be thought of as a set of complex numbers. Complex number notation is integral to description and analysis of APSK. Each symbol, x, is defined by a magnitude and a phase:[12]

$$x = |x|\angle x = |x|e^{i\angle x} = |x|\cos(\angle x) + i|x|\sin(\angle x) = x_R + ix_I$$

A set of symbols is then defined by a set of amplitudes, A, and a corresponding set of phases, θ. The actual value of A is usually irrelevant, and only the ratio between A_k is important.

$$x_k = A_k e^{i\theta_k}$$

The set of symbols is known as a constellation and is usually shown graphically as a set of points on a complex plane. Each symbol in the constellation maps to a set of data bits.[13] For example, the simplest PSK constellation, known as binary PSK or BPSK, consists of two symbols $\{-A, +A\}$, corresponding to data bits '1' and '0'. Equivalently BPSK is defined by: $A_0 = A_1 = A; \theta_0 = 0, \theta_1 = \pi$, and is shown in Fig. A.5a. More complex constellations, shown in Fig. A.5b–d, use multiple phases (4 for QPSK, 8 for 8PSK) and multiple phases and amplitudes (12 phases and 3 amplitudes for 16QAM). Special cases of the general constellation described above are described below:[14]

- If the phase is same for all symbols and only the amplitude changes, the constellation is known as amplitude shift keying, ASK. ASK is rarely used in modern digital communications.

---

[12] The real part of x, x_R, is frequently called the in-phase part (I) while the imaginary part, x_I, is called the quadrature part (Q).

[13] Any mapping is allowed, but Gray coded (i.e. binary representation of symbols that are close to each other differ by only one bit) maps usually result in the best performance.

[14] Note that some constellations fall under multiple special cases.

Fig. A.6 Differential BPSK

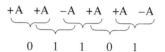

- If the amplitude is the same for all the symbols in a constellation and only the phase changes, the constellation is known as phase shift keying m-PSK.[15]. PSK with $m > 8$ is rarely used in modern digital communications since its performance is inferior to that of QAM.
- If the symbols are arranged on a rectangular grid, the constellation is known as mQAM.[16]
- If symbols have different amplitudes and are *not* arranged in a rectangular grid, the constellation is known as APSK.

A comprehensive review of QAM and QPSK is presented in [5]. The choice of amplitudes and phases has a significant effect on the performance of the communications link using that constellation. Constellation design is usually a tradeoff between desired throughput, noise immunity, and implementation complexity.[17] Throughput is related to the amount of information carried by each symbol, as discussed in Sect. A.1 ($\log_2(\# \text{ of symbols in constellation})$, which equal to 1 for BPSK, 2 for QPSK, 3 for 8PSK, and 4 for 16QAM).

A.4.5 Differential Modulation

Some modulations encode the information not in a single symbol, but in a sequence of symbols. The most common type of this modulation with memory is known as differentially encoded. In a differentially encoded modulation, the information is extracted based on the *difference* between two or more successive symbols. For example, a differential BPSK (DBPSK) modulation encodes a '0' by transmitting the same symbol as the previous symbol, while a '1' is encoded by transmitting a different symbol than the previous symbol, as shown in Fig. A.6. The same approach works with FSK, where the frequency difference between successive tones determines the transmitted bit sequence.

Differential modulation is often used when the absolute value of each symbol cannot be determined (e.g. the actual phase of a symbol is unknown). The previous symbol can then be used as a reference for the current symbol. The simpler

---

[15] Note that PSK is *not* a continuous phase modulation

[16] Note that BPSK and QPSK can be also called 2QAM and 4QAM. However, the convention is to use BPSK and QPSK.

[17] The standard PSK and QAM constellations with more than 4 points are known to be suboptimal for AWGN [276] and fading channels [275] (by a fraction of a dB). However, receivers for optimal constellations are more difficult to implement and performance improvement is negligible.

Fig. A.7 Pulse position
modulation

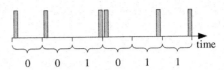

time

0   0   1   0   1   1

demodulation comes at the price of reduced performance. Corruption in one symbol affects two symbol decisions.[18] Double differential modulation uses three consecutive symbols instead of two [223] to provide even more immunity to frequency offset at a cost of higher degradation in performance.

A.4.6 Pulse Position Modulation

Pulse position modulation (PPM) relies on transmitting a short pulse at different times within a symbol period. Unlike the previously described modulations, the signal is transmitted for only a small portion of the symbol period.[19] For example, a pulse sent in the first half of the symbol period can be mapped to a data bit '0', while a pulse in the second half of the symbol period is mapped to '1.' An example is shown in Fig. A.7. Multiple bits can be transmitted by allowing multiple pulse positions within a symbol. One of the main difficulties in implementing a PPM is accurately determining the symbol period boundaries. Differential PPM (see Sect. A.4.5) is frequently used to avoid having to determine the symbol boundaries.

PPM is most often used for optical communications, both over fiber [224] and over the air [225]. RF transmissions using PPM are less common. Short duration of the burst makes the PPM waveform wideband (i.e. occupy a large bandwidth relative to the symbol rate) and has attracted some interest from the ultrawideband (UWB) community [226].

A.4.7 Orthogonal Frequency Division Multiplexing

OFDM is a popular modulation that has been adopted for well over half of all wideband terrestrial standards (including both contenders for fourth generation cellular telephony). This section barely scratches the surface of this sophisticated modulation, and the reader is encouraged to see Chap. 12 in [1] for a quick

---

[18] Actual performance degradation depends on the details of the modulation and demodulator implementation. In the simplest case, an uncoded, hard-decision, coherent D-BPSK demodulator generates twice as many errors as the non-differential demodulator. At high SNR this degradation is small (fraction of a dB), but at low SNR the degradation is up to 3 dB.

[19] Short burst transmissions can be very desirable for some hardware implementations. For example, a laser may be powered from a capacitor that takes a lot longer to charge than discharge. Short bursts result in a wideband signal, making PPM one of the modulations selected for the UWB standard [284].

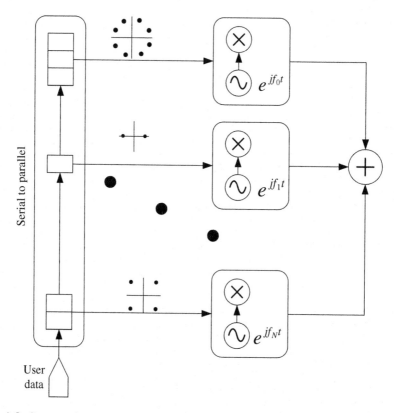

Fig. A.8 Conceptual structure of OFDM modulation ($f_k = 2\pi f_k$ for conciseness)

summary or Chap. 15 in [5]. This section first describes the modulation, and then explains why it is so popular.

An OFDM modulation consists of N subcarriers.[20] The serial user data stream is converted to a parallel vector with one or more bits for every used[21] subcarrier. The number of bits allocated to each subcarrier corresponds to the modulation selected for that subcarrier (e.g. 1 for BPSK, 3 for 8PSK). The modulation on each subcarrier can be different. The bits for each subcarrier are then mapped into a modulated symbol. The symbol modulates a subcarrier of frequency f_k (Fig. A.8). So far, this describes a standard FDM waveform (see Sect. A.5.2). The key to OFDM is setting $f_k = k \times f_b$, where f_b is equal to the symbol rate. This frequency assignment ensures that the subcarriers are packed tightly and still remain orthogonal to each other – no subcarrier interferes with any other subcarrier. The number of subcarriers in practical systems ranges from 64 to 4096. Clearly, a brute

---

[20] N is almost always a power of 2 to allow efficient hardware implementation using FFTs.

[21] Some subcarriers can be designated as 'unused' or 'pilots', or even allocated to other users in the case of OFDMA.

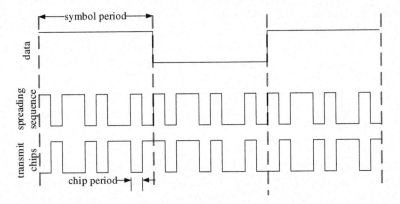

Fig. A.9 Spreading a signal by a high rate chipping sequence (SF = 10, PG = 10 dB)

force implementation of the block diagram in Fig. A.8 is impractical. All currently fielded OFDM systems use an FFT to efficiently multiply the input symbols by a set of complex exponentials. However, an FFT is not essential to an OFDM-like modulation as discussed in [227, 228].

The primary motivation for OFDM is its performance in a fading environment (see Sect. A.9.3). A wideband single carrier waveform experiences different channel gain at different frequencies in the band. This leads to distortion known as intersymbol interference (ISI). ISI caused by a wideband rapidly changing channel is difficult to mitigate. Each subcarrier in a wideband OFDM waveform 'sees' a small portion of the channel bandwidth and the channel gains is essentially constant over the frequencies in the subcarrier band. Thus, no ISI is introduced and no mitigation is necessary. This convenience comes at the price of a small amount of overhead due to the cyclic prefix.

A.4.8 Direct Sequence Spread Spectrum

DSSS is a modulation technique that aims to spread the transmitted signal power over a bandwidth much larger than the symbol rate. Since the bandwidth is roughly proportional to the rate of change in the signal, spreading is accomplished by multiplying the original signal by another signal. The second signal has a much higher symbol rate. The symbol rate of the second signal is known as the chip rate. The simplest implementation of a DSSS modulation is shown in Fig. A.9. The data signal modulated with BPSK is multiplied by a high rate signal, also modulated with BPSK. Each 'bit' in the high rate signal, known as a chip, is generated using a pseudo-random sequence. The final output has transitions at the chip rate and therefore occupies a bandwidth proportional to the chip rate. The ratio of the chip rate to the symbol rate is known as the spreading factor, SF.[22] The spreading factor is also expressed as the processing gain, PG $= 10 \log_{10}(SF)$.

The receiver multiplies the received signal by a replica of the pseudo-random code to get back the original data signal (multiplying each chip by itself effectively removes the spreading since $1 \times 1 = 1$ and $-1 \times -1 = 1$).

Since spectrum is a precious resource, why would it be desirable to *increase* the bandwidth of a signal? At first glance, it seems quite wasteful. There are three reasons to use DSSS:

1. **Security:** The total power in the spread and unspread signals is the same. Then the power spectral density (power per Hz of bandwidth) is lower for a spread signal. For large SF, the power spectral density is very low and can easily be lower than the noise floor (i.e. natural noise level in the environment). The receiver can still detect the signal by 'despreading' it. However, a hostile observer cannot easily detect that a transmission is even taking place,[23] leading to a low probability of detection (LPD). Even if the transmission is detected, the observer would have to know the pseudo-random sequence to make sense of the signal.

2. **Multiple Access:** Each spread signal appears as noise to other users in the *same* band. As long as the pseudo-random sequences for each user are relatively uncorrelated, multiple users can share the same band at the same time (also see Sect. A.5.5).

3. **Multipath Fading Mitigation**: As discussed in Sect. A.9.3, multipath fading is usually more detrimental to wideband signals than to narrowband ones. However, efficient receiver architecture exists to mitigate the effects of multipath fading on DSSS waveforms. A RAKE receiver (Chap. 13 in [1]) recombines delayed replicas of the signal in the time domain to combat fading.

DSSS modulation is used for all third generation cellular telephony standards (3GPP and cdma2000) as well as for some WiFi modes for reasons 2 and 3. It has been replaced by OFDM for the fourth generation of wireless standards.

Most examples of DSSS use either BPSK or QPSK for both data and chipping modulations. However, it is entirely valid to spread a BPSK data signal with an 8PSK chipping sequence.

A.4.9 Frequency Hop Spread Spectrum

FHSS is similar to DSSS in that it aims to spread the transmitted signal power over a bandwidth much larger than the symbol rate. However, instead of using a high-

---

[22] The chip edges are aligned with the symbol edges and SF is an integer in a synchronous DSSS waveform. The chipping rate is independent of the symbol rate (i.e. SF is not necessarily an integer) in an asynchronous DSSS system. Most modern DSSS systems are synchronous.

[23] Sophisticated techniques can detect a spread signal below the noise floor.

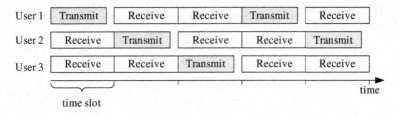

Fig. A.10 Simple TDMA schedule

frequency chipping sequence, the carrier frequency of the signal is varied over the spreading bandwidth. The carrier frequency is changed according to a predefined formula every hop period. A hop period can consist of many symbol periods (slow hopping) or a fraction of one symbol period (fast hopping). Frequency hopping was first used for military electronic countermeasures. Because radio communication occurs only for brief periods on a radio channel and the frequency hop channel numbers are only known to authorized receivers of the information, transmitted signals that use frequency hopping are difficult to detect or jam.

FHSS is usually easier to implement than DSSS. A popular consumer wireless standard, Bluetooth, selected FHSS because it can be easily implemented in a low-power device.

FHSS can be thought of as a special case of TDMA/FDMA multiple access scheme (see Sect. A.5.4) with a user assigned a different frequency band in each time slot.

A.5 Multiple Access

Multiple access techniques specify how different users can access the radio spectrum to communicate with each other. This section introduces some but not all of the commonly used techniques.

A.5.1 Time Division Multiple Access

In the simplest scenario, two users communicate using the same spectrum (same carrier frequency and bandwidth) by taking turns transmitting and receiving. This approach can be extended to more than two users by giving each user an exclusive opportunity to transmit at a given time. All other users must not transmit at the same time, as shown in Fig. A.10. The transmit opportunity is known as a time slot. TDMA requires that all users have the same concept of time and can time their transmissions accurately. Accurate timing can be accomplished using either:

Fig. A.11 Simple FDMA assignment

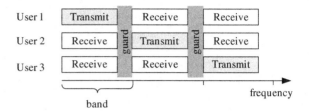

- A master/slave relationship. The master user (or a super-user, also known as a basestation) transmits a signal at regular intervals. All other users lock onto that transmission and set their internal clocks to it. The master user is typically also responsible for indicating which time slot(s) each user can employ for signal transmission.
- An external timing source (e.g. GPS) can be used to synchronize all the users. Some mechanism is still needed to assign time slots to users.

Absolutely accurate timing is never possible, and a small guard time must be inserted between successive transmissions. Sophisticated TDMA systems estimate the distance between users and take the propagation delay between them into account to adjust transmission start and stop times.

Slot assignment can be dynamic, with users requesting more or fewer time slots as their throughput requirements change. Note that TDMA does not allow for true full-duplex (i.e. simultaneous bidirectional) communications, but if the time slots are short enough, an appearance of full-duplex is created. TDMA provides a simple way to support variable throughput requirements for multiple users by assigning multiple time slots to the same user. For example, if user 2 is downloading a file while user 1 is only talking on the phone, user 2 may get three time slots for every one that user 1 gets.

A.5.2 Frequency Division Multiple Access

Users can be separated in frequency, with each users assigned a frequency that does not overlap with any other users' frequency. The frequency and bandwidth allocated to each user is known as a frequency band. Unlike time, which must be set from an external source, frequency is set relatively accurately by a crystal in the user equipment. Since the users do not interfere with each other, full duplex communication is possible. There is no need for a master user to set time slots. However, a master may be required to assign bands to users. FDMA is familiar to most readers since it is used to support multiple TV and radio channels. A simple frequency assignment is shown in Fig. A.11. Frequency uncertainty between different users requires a small guard band between each users' band. The guard band can be made smaller if all users synchronize to a common reference. The same synchronization techniques can be used for FDMA as for TDMA.

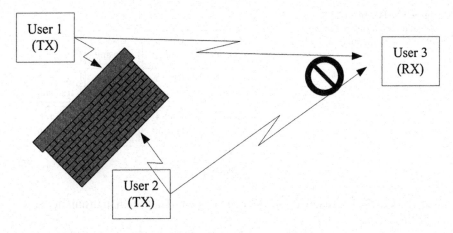

Fig. A.12 Hidden user problem with CSMA

FDMA makes it difficult to respond to changing throughput requirements. While it is possible to allocate multiple bands to a single user, it is not usually done[24] and all users always get the same throughput.

A.5.3 Random Access

Perhaps the simplest possible multiple access scheme allows all users to transmit whenever they have data available. Clearly, if two users transmit at the same time in the same area, the transmissions will collide and neither user succeeds in getting his data to the receiver. In that case, the users retransmit after a (typically random) delay. The throughput of such a system, known as pure ALOHA, is quite low if there are many users.

Carrier sense multiple access (CSMA) improves on pure ALOHA. At any given time the user checks if another transmission is already taking place (i.e. senses the channel). If the channel is clear, the user transmits his message. If the channel is busy, the user either continues sensing until the channel is free (known as persistent CSMA) or waits a random amount of time to try again. CSMA is very simple and requires no infrastructure but results in lower overall throughput due to collisions between users. CSMA is only applicable for packet-based[25] data rather than streaming data. CSMA has better throughput than ALOHA because the number of collisions is reduced. Sensing an ongoing transmission can be difficult and unreliable if a fading channel (see Sect. 3.4), or simply because two transmitters cannot 'see' each other, as shown in Fig. A.12. In that case, both users 1

---

[24] Except in OFDM and OFDMA systems, but those are not strictly speaking FDMA.

[25] Most modern wireless systems are packet-based.

Fig. A.13 TDMA/FDMA
slot assignment (users 1 and 2
get higher throughput)

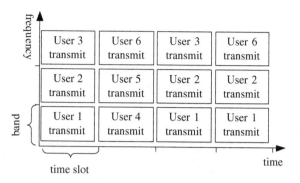

and 2 'sense' the channel and declare it open. They both transmit to user 3 resulting in a collision.

Collision avoidance extends CSMA (CSMA/CA) by having the receiver transmit a busy tone as soon as it starts processing a transmission. Since collisions occur at the receiver, the hidden user problem described above is somewhat alleviated. All transmitters close to the receiver now know not to transmit.

A.5.4 Time and Frequency Multiple Access

TDMA and FDMA can be combined to support more users. In the hybrid TDMA/ FDMA approach, each user is allocated a time slot and a frequency band it can transmit in, as shown in Fig. A.13. Taking advantage of both dimensions results in a more flexible schedule and more efficient use of the spectrum. For example, users can be allocated multiple timeslots and/or bands when they require higher throughput. The GSM standard uses this approach to support many cellular phones per basestation.

A.5.5 Code Division Multiple Access

CDMA is only applicable to DSSS modulations described in Sect. A.4.8. Each user is assigned a pseudo-random chipping sequence, PN_k. The sequences are designed such that the peak cross-correlation ($\star$) between any two length N sequences is approximately zero:

$$\max\left(PN_i \star PN_j\right) = \begin{cases} \varepsilon_{ij} \approx 0 & i \neq j \\ N & i = j \end{cases}$$

All users then transmit their spread sequences simultaneously in the same frequency band. Each user can extract the signal intended for him by correlating the sum of received signals with his chipping sequence. All other users' signals

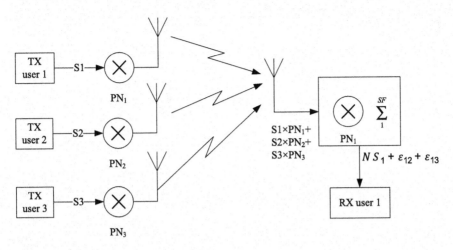

Fig. A.14 Code division multiple access

correlate to approximately zero and are treated as noise. An example, shown in Fig. A.14, demonstrates the concept for three users.

An obvious question is why an infinite number of signals cannot be transmitted this way in the same band. Theoretically, this is possible as long as the total throughput does not violate the channel capacity theorem (see Sect. A.3) extended to a multiple access channel (p. 474 in [229]). For relatively simple receivers,[26] the non-zero cross-correlation limits practical number of users that can share the bandwidth [230, 231].

A.5.6 Carrier-in-Carrier (Doubletalk) Multiple Access

A unique multiple access technique was patented about 10 years ago. It allows two users to share the *same* band at the *same* time without using DSSS. To date, this technique has only been applied to satellite communications and has very limited exposure in open literature [232, 233]. It is described here only because of the 'wow' factor.

The concept is illustrated in Fig. A.15. Two users cannot communicate directly with each other, and use a satellite to relay their signals. For the purpose of this example both users are in the same downlink beam from the satellite (i.e. both users receive whatever the satellite transmits). Users require full duplex communication, meaning they transmit at the same time. A conventional approach would be to use FDMA and allocate each user a frequency band in

---

[26] I.e. receivers that do not implement multi-user interference cancellation. Interference cancellation can reduce the effect of non-zero cross-correlation by subtracting the interfering users' signals from the received signal.

Fig. A.15 Carrier-in-Carrier (doubletalk) multiple access scheme: concept

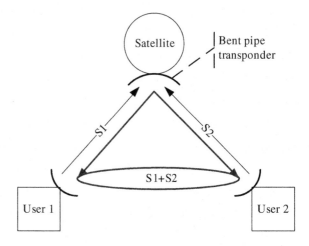

Fig. A.16 Carrier-in-Carrier (doubletalk) multiple access scheme: implementation

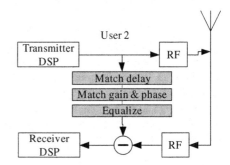

the satellite downlink. The total bandwidth used by the two users is the sum of their individual bandwidths. Satellite bandwidth is very expensive.[27] A basic loopback satellite transponder takes the composite signal received on its antenna, amplifies it, and transmits it on a different frequency. For the carrier-in-carrier technique, user 1 and 2 transmit their signals on the *same* frequency. The satellite then transmits the composite signal, S1 + S2 to both users as shown in Fig. A.15.

User 2 then has to find S1 from S1 + S2. In general, if S2 is not known, this is a very difficult or even impossible problem. However, user 2 knows exactly what S2 is since it just transmitted the signal. Therefore, all[28] user 2 has to do is (see Fig. A.16):

1. save a copy of the transmitted signal
2. delay it by the round trip delay to align it with the received signal

---

[27] Satellite capacity is leased by the end users in units of MHz × month. Actual numbers are closely held by the operating companies but are on the order of $5000/Mhz/month.

[28] As usual, the devil is in the details. Getting the phase and amplitude just right is difficult. If the relative amplitudes of S1 and S2 are very different (e.g. one is 100 times larger), compensation becomes even more difficult.

Fig. A.17 Carrier-in-carrier multiple access scheme without a satellite

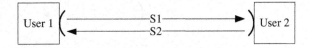

3. compensate for gain and phase changes introduced during the trip
4. subtract the delayed and adjusted copy from the received signal.

The astute reader will no doubt wonder why the same technique is not used without a satellite in the middle. Why does the approach shown in Fig. A.17 not work? Why can not User 2 simultaneously transmit and receive? Theoretically, the same cancellation technique described above should work. In practice, the enormous difference between the power of the transmitted signal and the received signal makes interference cancellation very difficult or impossible. The receiver ADC has to sample S1 + S2, where S1 is easily a ten billion (100 dB) times stronger[29] than S1. The main problem is that the output of the power amplifier is not exactly equal to the signal generated in the transmitter DSP. The thermal and quantization noise present at the input to the amplifier gets scaled together with the signal. The amplified noise is much higher than the receiver noise floor and cannot be cancelled out. It may be possible to create a very low noise power amplifier (perhaps a cryogenic one), but such a system is not practical today. Another option is to put in a high isolation circulator that prevents transmitted signals from feeding back to the receiver. Conventional circulators provide about 20 dB of isolation, which is not nearly enough. However, exotic photonic circulators have been reported that may offer sufficient isolation [167].

A.6 MIMO

A relatively new technology, known as multiple-input multiple-output (MIMO), uses multiple antennas at the transmitter and/or receiver to achieve higher throughput in the same amount of bandwidth. MIMO is a very large and complex topic and is the subject of many books. In this section, we briefly mention some of the benefits of MIMO and a qualitative argument for why it works. There are three fundamental methods to take advantage of multiple antennas:

1. Scale and time-align signals for each antenna to form a beam that enhances the desired signal and suppresses undesired signals. This technique is known as *beamforming* (see Sect. 10.5) and has been used for decades.
2. Combat effects of fading by relying on *diversity*. If each antenna sees an 'independent' version of the transmitted signal, it is unlikely that all the antennas will experience a 'fade' at the same time. Diversity gain is perhaps the

---

[29] For example, a signal transmitted at 2 GHz frequency over 1 km range experiences free space loss of ∼ 100 dB.

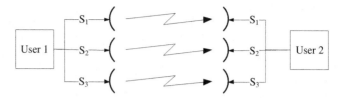

Fig. A.18 Simple MIMO system using highly directional antennas

most important application of MIMO for terrestrial wireless communications.
3. Increase capacity by sending different data on each antenna. This technique, known as *spatial multiplexing*, is the most recent application of MIMO.

The first two methods are very effective in the low-SNR regime, while the third method is most effective in the high-SNR regime. For all of these methods, a MIMO system must have two or more antennas at the receiver, transmitter, or both. The reader is likely familiar with the first two methods, and only spatial multiplexing will be discussed below.

The most fundamental results in MIMO were derived by Foscini [234] and Telatar [235]. They showed that the capacity of a channel using M_t transmit and M_r receive antennas increases **linearly** with max (M_t, M_r). This is a dramatic improvement over the logarithmic growth described in the previous section. MIMO systems achieve capacity that is at first glance higher than Shannon capacity by creating multiple parallel channels instead of just one. MIMO works best in an environment with many multipath. Each path provides another 'view' of the transmitted signals. Ideally, each antenna should receive a signal uncorrelated to any other antenna. In a rich multipath environment this can be achieved by simply separating antennas by at least ¼ of the wavelength of the signal.[30] The receiver then has many uncorrelated copies of the different transmitted signals and can apply signal processing to extract each signal.

A simple, but illustrative example is shown in Fig. A.18, where three highly directional antennas are used to transmit three separate data streams.[31] The beam from each antenna does not interfere with other beams. As far as the user is concerned, each antenna has access to the entire spectrum allocated to him, *B*. The total bandwidth is therefore *3 × B*. However, the total power available to the user does not change and must now be shared between the three antennas. For simplicity, assume that the signal from each antenna experiences the same

---

[30] MIMO is rarely used at low frequencies because the wavelength is large. Smaller wavelength allows for smaller antennas and less separation between them. For example, at 3 GHz, the wavelength is 10 cm. Antennas should be separated by at least 2.5 cm. A typical handheld device can host not more than two antennas.

[31] This technique (but without the restriction of directional antennas) is known as VBLAST [277]. The antennas are normally not directional. A 'directional' link is created by appropriately combining signals from all antennas. A matrix defined by the multipath coefficients is inverted and used to multiply the vector of received samples.

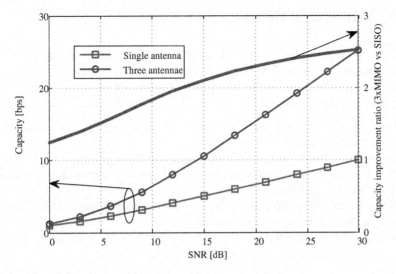

Fig. A.19 Capacity for a 3×3 MIMO system with uniform SNR

attenuation and the SNR at all receive antennas is the same.[32] The total power is
then shared equally between the antennas, $P_k = P/3$. The link capacity is then:

$$C = \sum_{1}^{3} B \log_2 \left(1 + \frac{P/3}{N} \right) = 3B \log_2 \left(1 + \frac{P/3}{N} \right)$$

Relative capacity for $B = 1\ Hz$, for the 3×3 antenna MIMO vs. a single
antenna (SISO) is shown in Fig. A.19. As can be seen from that figure, the MIMO
system achieves almost $3\times$ higher capacity than a SISO system at high SNR. The
last point is very important—spatial multiplexing becomes effectively only at
relatively high SNR, when the SISO system is in a bandwidth-limited regime (see
Sect. A.3). Since MIMO is extremely popular, spatial multiplexing has been
suggested for systems that operate in low SNR and do not experience fading
(e.g. satellite links), where the capacity improvement cannot be realized.[33] MIMO
is applicable for satellite-to-mobile links that experience some fading [236].

SDRs are ideally suited for taking advantage of multiple antennas over a wide
range of channel conditions by choosing the appropriate algorithm for different SNR
regimes. For example, a SDR can enable full VBLAST processing when average link
SNR is in the 10 dB range. Once SNR falls below 10 dB, additional antennas can be
disabled to save power, or switch to beamforming to achieve power gain.

---

[32] This is almost never the case and power is allocated to different antennas using the waterfilling
algorithm. The waterfilling algorithm can be shown to be optimal for uncoded systems, but is
suboptimal for coded systems [289].

[33] For example, it does not make sense to use capacity improving MIMO when using a powerful
code and a robust modulation such as one shown in Figure A.22.

A.7 Performance Metrics

The existence of dozens of different modulations and wireless standards indicates that there is no single 'best' waveform. Each modulation has advantages and disadvantages along a set of different metrics. The metrics are given different weight depending on the system requirements and constraints. A few relevant metrics are:

- **Bit error rate performance in the presence of noise**: What percentage of bits is received correctly?
- **Bandwidth efficiency**: What throughput can be achieved in a given bandwidth. The theoretical limit is given by the channel capacity, but most waveforms are quite far from that limit.
- **Implementation complexity**: Can the radio implementing the waveform be built within the technology constraints.
- **Robustness to interferers**: How well will the waveform perform in the presence of intentional or unintentional interferers.

A.7.1 Bit Error Rate

BER is perhaps the most fundamental, and definitely the most frequently used, metric applied to wireless waveforms. BER curves show the average fraction of bits received incorrectly versus the energy expended to transmit one bit of data relative to the noise floor, E_b/N_0. E_b/N_0 is closely related to the signal to noise ratio.[34] BER curves are also known as waterfall curves because of their characteristic shape, see Fig. A.20. Since each curve captures only the dependence on SNR, all other relevant parameters (e.g. channel type, interference environment) must be annotated separately. The BER curve can be used by the system designers to determine the required transmit power, or the distance over which the link can operate by computing these parameters from E_b/N_0. Analytical equations have been derived to compute the BER for almost every uncoded waveform and for a small subset of coded waveforms. Most of these equations are only applicable in a pure AWGN channel, and some are available for fading channels.[35]

Analytical results are invaluable for providing intuition about what variables affect the BER. However, these results are derived for simplified and idealized operating environments which are rarely seen in practice. BER curves for a specific scenario that takes into account all the distortions and degradations are

---

[34] The author has sat through many meetings where the exact meaning of the abscissa was debated over and over. Some researchers choose to use the *symbol* rather than bit power, E_s/N_0, which are not the same if each symbol conveys multiple bits. Others feel that the forward error correction coding should not be considered in the definition of E_b/N_0, and use the term 'channel bit', E_c/N_0. Still others are only comfortable with 'true' SNR, or C/N_0. Each community has its own preferences (e.g. GNSS prefers C/N_0).

[35] MATLAB provides a nice compendium of these analytical results via the 'bertool' command.

Fig. A.20 A few BER curves

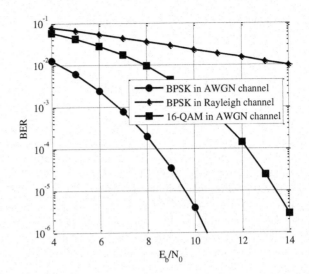

usually obtained by simulating the system. These simulations can be extremely computationally intensive and require a long time to complete. One of the applications of SDR is the simulation of BER curves. Hardware-based SDRs can speed up these simulations. Of course, the SDR has to implement all the relevant impairments and degradations. The difference between the simulated curves and the analytical ones is known as performance loss.

BER curves are used almost universally to compare performance of different waveforms. However, it is often the wrong metric. Uniform distribution of errors is an implicit assumption in a BER curve. Consider two ways to get BER = 10 % in a set of 100 bits:

1. 9 correct bits, followed by 1 incorrect bit. The sequence repeats 10 times.
2. 90 correct bits followed by 10 incorrect bits.

The first scenario is likely to occur in an AWGN channel with an uncoded waveform. The second scenario can occur in a fading channel, or if coding is applied. The impact on the user from these scenarios can be quite different:

a. Live video is transmitted with 10 bits per video frame. In the first scenario, each frame has a minor glitch that may not be noticeable. In the second scenario, an entire frame is lost which is quite noticeable. Thus, scenario 1 is better than scenario 2.
b. Data is transmitted in packets of 10 bits. Parity is checked on every packet and if the check fails, the packet is discarded. In the first scenario, all packets contain one error and therefore all are discarded. The packet error rate is 100 %. In the second scenario, only the last packet is discarded and the packet error rate is 10%. Thus, scenario 2 is better than scenario 1.

Packet error rate is a metric more suitable for modern wireless systems. It does not matter how many bits in a packet are bad—one invalidates the entire packet.

An interesting example of an engineering error due to reliance on BER is described in [237].

A.8 Forward Error Correction

Forward error correction is a technique that adds redundant bits to the transmitted data to help the receiver correct errors. The algorithm for computing these redundant bits is known as the error correcting code. The channel capacity equation described in Sect. A.3 states that there exists an FEC code that achieves capacity while maintaining an essentially zero BER. Unfortunately, Shannon provided little practical guidance to actually design such a code.[36]

Before looking at FEC, let us consider how well an uncoded modulation performs relative to channel capacity. For the sake of argument, let BER of 10^{-6} satisfy the 'essentially zero' requirement. Then, looking at Fig. A.20, we see that E_b/N_0 of ~ 10 dB is required to achieve it for a BPSK waveform in AWGN channel. Recall that bandwidth is proportional to the symbol rate,[37] and that BPSK transmits one bit per symbol. Then, at 10 dB of SNR, BPSK achieves a throughput of 1 bps/Hz. The Shannon capacity is $\log_2(1 + 10) \approx 3.5$. At 10 dB SNR, BPSK is $3.5\times$ below theoretical capacity. Another way to look at it is that theoretical capacity of 1 bps/Hz is achieved at SNR of 0 dB ($\log_2(1 + 1) = 1$). Uncoded BPSK requires $10\times$ more power than 'it should'. How can we close this gap?

Error correction codes differ in the way the redundant (parity) bits are computed. The more powerful codes require fewer redundant bits, K, to correct more bit errors. A key parameter in any FEC code is the code rate,[38] the ratio of original data bits, N, to the number of transmitted bits, $N + K$.

$$r = \frac{N}{N + K}$$

In general, codes with lower rates can correct relatively more errors and can therefore work at lower SNR. Dozens of code families have been developed over the past six decades, and each family includes a number of variations. Theoretical BER performance has been derived for some of the older and simpler codes (e.g. Reed-Solomon and Viterbi), but is not (yet) known for the newer codes (e.g. Turbo, LDPC).

---

[36] Very long 'random codes' are one way to achieve capacity, but these codes cannot be practically encoded or decoded.

[37] For simplicity, assume the bandwidth is equal to the symbol rate. In a practical system the bandwidth is always larger than the symbol rate, making the results in this paragraph even worse.

[38] A class of rateless codes does not have a fixed r. Instead, the number of parity bits is essentially unlimited and the encoder continues to transmit parity bits until the decoder indicates that it has successfully received the data [280].

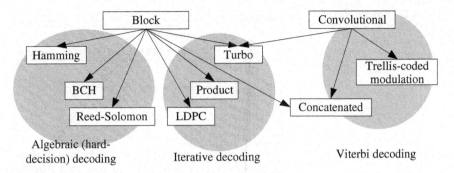

Fig. A.21 Classification of FEC codes

Codes can be classified into two major classes, as shown in Fig. A.21:

1. Block codes work on fixed length sequences of bits. The parity bits depend only on the previous N bits. Code performance generally improves (within a code family) as the block size increases. Examples of block codes include: Reed-Solomon, Turbo, LDPC. Most modern powerful codes are block codes.
2. Convolutional codes work on sequences of arbitrary length. A feedback loop may exist in the equation for computing parity bits and the parity bits then depend on the entire history of transmitted bits. Since most modern systems are packet-based, convolutional codes are typically terminated to keep the history a fixed length. The best known convolutional code is actually named for the type of decoder used to process it – the Viterbi decoder (frequently, but incorrectly, called the Viterbi code).

FEC decoders can be classified into two major classes:

1. Hard-decision decoders work on data bits estimated from the received symbols.
2. Soft-decision decoders work on real values that indicate the reliability of the data bits estimates from the received symbols. For example: BPSK symbols corresponding to bits '0' and '1' are nominally +1 and −1. Symbols are received corrupted by noise and are digitized to real values {0.7, -0.99, 1.2, 0.2}. The hard-decisions are then {'0', '1', '0', '0'}. Clearly some information is lost going from the received symbols to hard decisions. Soft-decision decoders take advantage of this 'lost' information. For example, it is clear that the confidence that the last symbol is a '0' is a lot lower than for the previous symbol (0.2 is a lot closer to −1 than 1.2 is).

Soft-decision decoders provide better performance (by as much as 3 dB, depending on the code), but require more computational resources (at least 4×). A major advance in FEC code design came in the 1980s with the development of iterative decoding for Turbo Codes. Prior to iterative decoding, a decoding algorithm was applied to the received data once. Whatever errors remained after the decoding algorithm could not be corrected. The elegant idea of iterative decoding is to apply the same algorithm multiple times. For specially designed codes, each application of the decoding algorithm changes the reliability estimates

Fig. A.22 Performance of coded (DVB-S2 r = ½ 64 kb block size) BPSK vs. uncoded

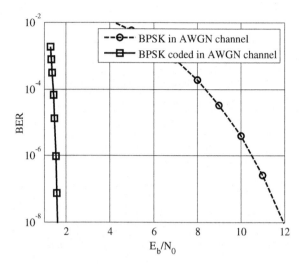

for the data bits.[39] The information from data bits with higher confidence is used to increase the confidence in other data bits. After a number of iterations, all confidences are expected to converge to large values and decoding is complete. Applying the algorithm many times increases the computational complexity. An FEC decoder for a modern code such as Turbo and LDPC often dominates the computational complexity of the entire modem.

Figure A.22 shows BER performance of a powerful modern code (DVB-S2 LDPC, r = ½, N = 32kb). The difference between the coded curve and the uncoded curve is dramatic. The difference between E_b/N_0 required to achieve a certain BER for coded and uncoded waveforms is known as the coding gain. For example, the coding gain at BER $= 10^{-6}$ for this code is ~ 9 dB. This gain does come at the price of reduced bandwidth efficiency—using a rate ½ code reduces the throughput in a given bandwidth by a factor of 2. Keeping the same assumptions as in the example above, the throughput achieved by the coded waveform is 0.5 bps/Hz. Theoretical capacity of 0.5 bps/Hz is achieved at SNR of −0.3 dB ($\log_2(1 − 0.3) \approx 0.5$). The coded waveform requires about 2 dB more power than theoretical minimum, which is a lot better than 10 dB more for the uncoded waveform.

The coded BER curve is an almost vertical line, very different from the relatively smooth uncoded curve. Powerful modern codes turn the waterfall curves into cliffs. This is an important observation that is sometimes overlooked by engineers used to the uncoded systems. In particular, it means that the system does not degrade gracefully as SNR decreases. Old, uncoded waveforms, could be relied on to operate with somewhat degraded performance if the SNR dropped below the target. Modern waveforms fail completely.

---

[39] Most iterative decoders are also soft-decision decoders. However, this is not necessary and hard-decision iterative decoders exist.

A.9 Distortions and Impairments

"Noise is a communications engineer's best friend. Without it we would not have a job"
f. harris

A radio sends its signal into a hostile environment, bent upon degrading and destroying the signal before it reaches the intended receiver. The complexity of this environment, known as the channel, provides employment to most communications engineers. In addition to impairments introduced by the environment, the receiver and transmitter both distort the signal from the theoretical ideal. In this section, we review some of the most frequently encountered channel impairments. The first few subsections cover environmental effects, while the last few cover distortions introduced by the hardware.

A.9.1 Thermal Noise

The most fundamental impairment is thermal noise. Thermal noise is usually assumed to be a pure random process with Gaussian distributed amplitude and flat (white) spectrum. It is therefore called additive white Gaussian noise (AWGN). AWGN comes from many sources, from the vibration of atoms in the receiver electronics, to black body radiation from the warm ground, to the background radiation left over from the creation of the universe. Thermal noise exists at all frequencies (at least from DC to 10^{12} Hz [2], p. 31) at a constant spectral power density of N_0 dBm/Hz.[40]

The total noise power (variance of the random process) is given by $P = k \times T \times B$, where T is the temperature[41] in Kelvin, B is the bandwidth in Hz, and k is Boltzmann's constant. At room temperature, $N_0 = -174$ dBm/Hz, a number ingrained in the mind of every communications engineer.

Systems with only AWGN are analytically tractable and a vast majority of published results assume AWGN. Unfortunately, such systems are few and far between. Space communications with highly directional antennas and point-to-point microwave links are some of the examples.[42] The channel for terrestrial communications such as cellular telephony is considerably more complicated.

---

[40] Note that the total power in thermal noise appears to be infinite (constant power density × infinite bandwidth). In fact, noise power does decrease at very high frequencies, but is white for all practical purposes for wireless systems. High total power is one reason why it is difficult to build a good noise source for laboratory testing. Putting wideband noise into a wideband amplifier can overdrive that amplifier and introduce nonlinear effects. This unexpected effect is often overlooked even by experienced engineers.

[41] Cryogenics is used to lower the temperature of the front end (low noise amplifier) of ultra-sensitive radios [262].

[42] Even these examples are AWGN only to a first degree – second-order effects such as rain and other atmospheric distortions must be considered.

A.9.2 Path Loss

As the electromagnetic wave emitted from the transmitter antenna spreads out, the power at any given point decreases. This effect is obvious to anyone who has used the flashlight—objects closer to the light are brighter. In free space (i.e. line of sight in a vacuum), power decreases with the square of the distance. The ratio of received power, P_R, to transmitted power, P_T is given by

$$\frac{P_R}{P_T} \propto \frac{1}{d^2 f^2}$$

The proportionality constant depends on the specifics of the transmitter and receiver antennas. The path loss is greater at higher frequencies. The quadratic path loss is useful for analytical derivations and can be assumed as a first order estimate for line of sight terrestrial communications. However, it is a gross oversimplification. Signals at some frequencies (e.g. <2 GHz) propagate through air almost as well as through vacuum, while at other frequencies (e.g. 60 GHz) the atmosphere absorbs most of the signal power due to water and oxygen molecule resonance. More importantly, most terrestrial communications links are characterized by obstructions (e.g. walls) and reflections. Measurements in an urban environment ([2], p. 953) show that path loss is not really quadratic in distance, but the exponent varies from 1.5 to 4, with a mean of about 2.7.

A.9.3 Multipath

As the electromagnetic wave emitted from the transmitter antenna makes its way to the receiver, the wave reflects off different objects in the path. Each reflection may eventually arrive at the receiver with a different power and at a different time. This propagation environment is appropriately called multipath as shown in Fig. A.23.

The effects[43] of multipath propagation depend on the symbol rate and carrier frequency of the transmitted signal. An excellent in-depth treatment of the multipath channel is provided in [238] and [239]. Low frequency signals do not reflect well off most objects and few multipath are generated. A multipath channel with M paths is described by:

$$y(t) = \sum_{m=1}^{M} h_m x(t - d_m)$$

where d_m is the delay of the mth path and h_m is the gain of that path.

---

[43] One effect familiar to those old enough to have used analog TV sets is 'ghosting.' A faint copy of an older image was visible over the normal picture.

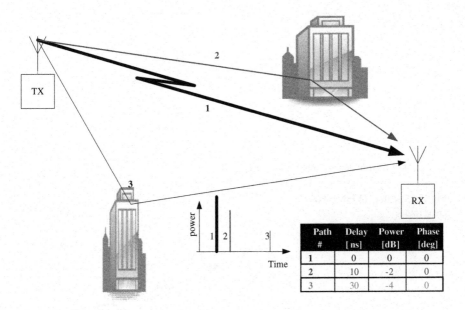

Path #	Delay [ns]	Power [dB]	Phase [deg]
1	0	0	0
2	10	-2	0
3	30	-4	0

Fig. A.23 Multipath propagation

The gain is usually expressed relative to the strongest path. The gain is typically complex-valued (i.e. non zero phase), but will be assumed real for the examples below. The multipath coefficients are usually not known *a priori* and change over time if the transmitter, the receiver, or anything in the environment is mobile. Multipath propagation has two effects on the received signal:

1. Inter-symbol interference
2. Fading

A.9.3.1 Inter-Symbol Interference

As delayed and attenuated copies of the transmitted signal add up at the receiver antenna, the resultant signal is a distorted version of the transmitted one. An example, assuming a BPSK signal at a symbol rate of 10^6, is shown in Fig. A.24. The distortion gets worse at higher symbol rates, where the delay between the paths is on the order of (or large than) a symbol duration. The 'typical' delay between paths depends on the environment and can vary from 100ns to 10μs or even larger. Inter-symbol interference is not a major concern for data rates below 10 ksymbols/s, but must be considered for higher data rates. Left unmitigated, ISI dominates overall system performance and can preclude

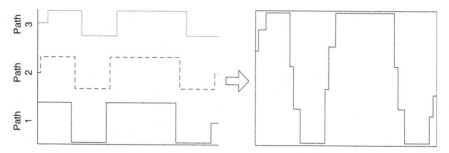

Fig. A.24 Effect of inter-symbol interference

Fig. A.25 Effect of ISI on bit error rate

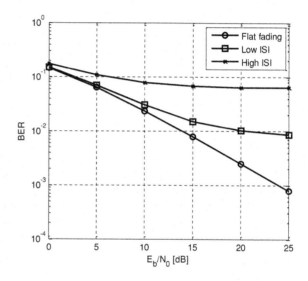

communications at data rates above 100 kbps even at very high SNR (Fig. A.25[44,45] compare to Fig. A.20).

A number of ISI mitigation approaches have been developed over the years:

Adaptive equalization. The multipath channel is a linear system and an approximate inverse linear system can be designed to cancel the effect of the channel. The problem can be stated as: find a set of filter coefficients z_k such that:

$$z(n) = \sum_{l=1}^{L} z_l y(n - lT) \approx x(n)$$

---

[44] This figure assumes a simple model for multipath – path power decreases linearly with the path delay. Low ISI assumes 3 total paths, while high ISI assumes 201 total paths. Note the 'error floor' behavior due to ISI.

[45] BER curves in a fading channel no longer have the 'waterfall' shape, but are closer to straight lines.

where T is the sample rate of the digitized signal. T is usually equal to half of the symbol period, $T = \frac{1}{2}T_b$. Adaptive signal processing techniques such as steepest descent can be applied to minimize $|x(n) - z(n)|$. Equalization works well as long as the maximum delay is not more than a few symbol periods and the multipath coefficients change slowly relative to the symbol rate. Unfortunately, neither requirement is met for modern high-rate systems that are expected to operate while the user is in a moving vehicle.

Direct sequence spread spectrum (see Sect. A.4.8) is another approach to mitigate the effects of multipath. The idea is to *increase* the apparent symbol rate by converting each data symbol into a sequence of shorter chips. The chip duration is designed such that delays between paths are larger than the chip rate, i.e. $d_m - d_{m-1} > T_c$. Then each path is treated as a unique signal, demodulated, phase and timing tracked. The individual signals are then coherently combined by appropriately scaling and delaying the symbols. A receiver that implements such an algorithm is known as RAKE. RAKE receivers do not work well if two or more paths arrive within a chip duration.

Orthogonal frequency division multiplexing (see Sect. A.4.7) takes an opposite approach to multipath mitigation by making the symbols much longer. An OFDM symbol consists of many low-rate subcarriers packed closely in frequency. Multipath will corrupt the beginning (d_M seconds) of the long symbol, but most of the symbol is untouched. The key idea behind OFDM is to take samples from the end of the symbol and *prepend* them to the beginning of the symbol. The prepended samples are known as a cyclic prefix (CP). Cyclic prefix must be longer than d_M. The FFT algorithm used to demodulate OFDM is circular, and any set of sequential samples can be used for demodulation. The multipath delay, d_M, is estimated and the first samples are discarded as shown in Fig. A.26.

A.9.3.2 Frequency Selective Fading

The electromagnetic waves from each path combine at the receiver either constructively or destructively. Fading is best understood by considering the high-frequency carrier rather than the data symbols. The carrier is a sine wave propagating in space. A delay is then equivalent to a phase shift. A delay of half a period (equivalently half a wavelength) is equivalent to a 180° shift, or a sign inversion. The wavelength depends on the carrier frequency as $\lambda = c/f$, where c is the speed of light. At a carrier frequency of 2 GHz, a wavelength is 15 cm. A difference in path delay of 0.5 ns (7 cm) results in a phase shift of 180°. Consider the effect of two paths with path delay equal to half a wavelength as shown in Fig. A.27. The two waves combine destructively as shown in Fig. A.28.[46] No amount of signal processing in the receiver can compensate for the lack of input signal at the antenna. However, moving either TX or RX a few inches fixes the

---

[46] Note that there is almost no ISI in this scenario since the delay between paths is negligible relative to the *symbol* period.

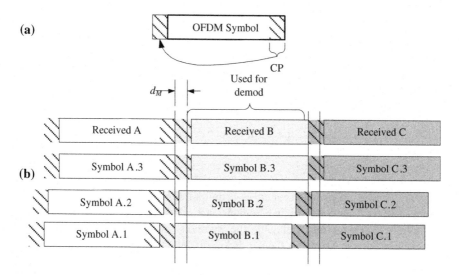

Fig. A.26 Using cyclic prefix in OFDM to mitigate inter-symbol interference (index .1, 2, 3 refers to path number)

Fig. A.27 Two path channel

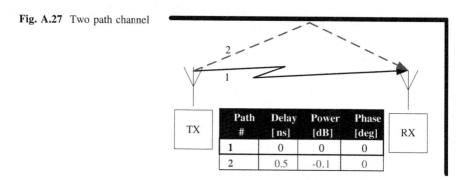

problem. This type of fading is known as fast fading because it changes rapidly over time if TX or RX is moving.

Wideband signals can be thought of as a many adjacent tones (that is literally the case for OFDM). Tones of different frequency end up with different phase shifts after the same delay and therefore combine to different powers. This is known as frequency selective fading, where parts of the signal spectrum are affected differently than others. Many mitigation techniques have been developed to combat fast fading.

Adaptive equalization is not particularly effective against fast fading because it cannot adapt quickly enough.

A combination of interleaving and forward error correction is effective in mitigating fast fading, if the fade duration is significantly shorter than the span of the interleaver. The symbols lost during a fade are randomly spread across the interleaver and then fixed by the FEC decoder.

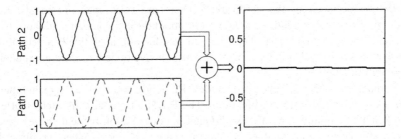

Fig. A.28 Destructive interference due to multipath

DSSS waveforms rely on the spreading gain to compensate for 'missing' parts of the signal. Intuitively, if the spread signal is 10 MHz wide, and 1 MHz of that is faded, only 10 % of the energy is lost and the signal can still be processed.

OFDM can use channel state information (if available) and put no data on the subcarriers that fall within the faded frequencies.

The most effective technique, known as antenna diversity, is to use multiple antennas. Consider a receiver with two antennas separated by at least half a wavelength. The probability that both antennas experience a deep fade at the same time is relatively low. More antennas reduce the probability even further. A simple multi-antenna receiver just selects the antenna with the highest received power and processes that signal. More sophisticated receivers take advantage of all the antennas and optimally combine their signals (e.g. using maximal ratio combining, MRC [238], p. 312). Many texts mention that MRC increases the *average* SNR of the link by a factor equal to the number of receiver antennas. While that is true, the effect is relatively unimportant compared to the dramatically reduced number of deep fades (and potential dropped links).

Frequency selective fading can also be caused by atmospheric scintillation. This effect is only relevant for ground-spacecraft links. Scintillation refers to the rapid fluctuation in signal amplitude and phase [240, 241]. The effects are often observed in satellite links and may occur either in the ionosphere due to changes in free electron density or/and in the atmosphere due to changes in pressure, humidity, temperature, etc. Although the mechanism behind scintillation is very different from multipath, the effect is very similar. An early example demonstrating the use of SDR to mitigate scintillation effects is described in [242].

A.9.3.3 Statistical Fading Channel Models

The number, magnitude, phase, and delay of multipath all depend on the geometry of the environment between the transmitter and receiver. It is clearly not practical to evaluate the performance of a radio for all possible combinations of geometric configurations. A statistical model of a channel is used to represent 'typical' environments. Such a model is used to generate time-varying channel coefficients, h_m, that can be used to first simulate the performance of radio algorithms and then verify performance of the hardware.[47] There are two types of models: those derived from highly simplified assumptions about the environment and those based on fits to measurement data. The 'derived' models are analytically tractable and are used in most academic publications. An extensive review of different models is available in [238] .

Perhaps the most widely used model, Rayleigh, assumes a large number of multipaths arriving from all directions. The Rayleigh model is assumes that each coefficient is a complex number with Gaussian distribution for real and imaginary parts. This model is applicable for dense urban environments where no direct line of sight exists between the transmitter and the receiver. The rate at which the coefficients change depends on the carrier frequency and the velocity of the radios. The coefficients are considered uncorrelated over time spans that exceed the channel coherence time. If a line-of-sight exists between the Tx and Rx, the Rayleigh model becomes a Rician model.[48] The Rician model adds a constant coefficient with zero delay.

A.9.4 Rain Attenuation

In addition to solid obstructions such as walls and trees, electromagnetic waves also have to contend with propagating through rain. Rain attenuation is of great concern to satellite communications because the signal has to go through the depth of the atmosphere filled with rain drops.[49] It is a smaller, but non-negligible, problem for long distance microwave links. The amount of attenuation, A, depends on rain intensity, R, and the signal carrier frequency, f. An empirical relation, $A = aR^b$, can be used to estimate the attenuation with coefficients $a(f)$ and $b(f)$ derived from measurement data (Fig. A.29) [243, 244, 245]. When the carrier wavelength is small, the waves get reflected and refracted by the individual raindrops. The effect is most pronounced for carrier frequencies above 10 GHz (wavelength <3 cm) [246]. Rain fading is a relatively slow process since it does not usually start raining hard immediately. ACM and power control are effective at mitigating the effects of rain fading.

---

[47] Channel emulators are available from many test equipment vendors (e.g. http://www.elektrobit.com/). These emulators typically take in an RF signal from the transmitter and output an RF signal to the receiver. Internally, a channel emulator uses one of the statistical models (or measured data) to filter the digitized input signal. A mid-range emulator runs $200,000 with MIMO models well over $500,000.

[48] Equivalently, a Rayleigh model is a special case of the Rician model with the ratio of the LOS power to the multipath power equal to zero. Note that both models are named after the equation that describes the distribution of received signal amplitude.

[49] The effect is no doubt well-known to readers with satellite TV receivers.

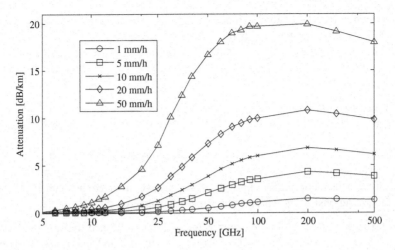

Fig. A.29 Rain attenuation as a function of carrier frequency and rain intensity

A.9.5 Interference

The vast majority of academic papers on wireless communications consider AWGN as the only source of interference. Real-world radios have to contend with many other sources of interference. In fact, most wireless networks are dominated by co-channel and adjacent-channel interference. For CDMA waveforms, imperfect timing alignment between different users causes non-zero correlation of different PN codes. FDMA waveforms suffer from power leaking in from adjacent frequency channels. In fact, almost every electromagnetic emitter is considered interference by some wireless network. Microwave ovens interfere with devices operating in the unlicensed ISM band. Power tools are notorious for spewing interference over a wide band. Military radios have to operate with intentional interference, known as jamming.

The main difference between AWGN and interference is that interference is typically not white (i.e. samples are correlated with each other). A SDR can take advantage of this correlation to mitigate some of the interference. Mitigation techniques are most effective if the interference is highly correlated but is uncorrelated with the signal of interest.[50]

SDRs excel at operating in difficult interference environments. Once the interference type is identified, an appropriate mitigation approach can be executed. Some of the approaches are:

- Frequency domain excision is effective against narrowband interference. The frequencies corresponding to the interferer are removed in either time or frequency domain.

---

[50] A type of jammer known as a 'repeater' [285] is very effective and requires a lot less power to achieve the same effect than a wideband noise jammer.

- An adaptive notch filter is tuned to optimally remove the interferer [247] or
- The received signal is transformed to the frequency domain using an FFT. The bins corresponding to the interferer are clipped or zeroed out. The frequency domain is then transformed back to the time domain for normal processing [248].

- Time domain blanking is effective against wideband bursty interference.
- Changing the waveform from DSSS to FHSS is effective against a repeater jammer.
- Changing the waveform from FHSS to DSSS is effective against a frequency follower jammer.
- Multi-user detection (also known as successive interference cancellation) is effective against a known and uncorrelated interferer.

Unexpected interference types are often encountered once the radio is fielded. SDR allows the waveforms and algorithms to be updated once new mitigation techniques are developed.

A.9.6 Nonlinearity

Transceiver hardware consists of many analog and RF components that introduce nonlinearities into the system.[51] Nonlinear distortion is difficult to mitigate and is avoided whenever possible. However, avoiding nonlinear distortion means that all the components have to operate in their linear regions. Many components, especially amplifiers, are most power efficient outside their linear region.[52] A trade-off exists between power consumption and the amount of distortion. The effects of distortion can be seen in both time and frequency domains. In the time domain, the effect is most obvious when looking at constellation plot, as shown in Fig. A.30a. The outer points (points with highest power) are compressed. Many wireless standards limit the level of distortion in the transmitter amplifier by specifying the maximum error-vector-magnitude (EVM). EVM is a measure of how much the distorted constellation differs from an ideal one. In the frequency domain, the spectrum becomes distorted due to intermodulation. A pulse-shaped spectrum acquires characteristic "shoulders" shown in Fig. A.30b.

Adaptive algorithms such as predistortion at the transmitter [249] and adaptive LLR computation at the receiver can be used to mitigate the distortion. Intermodulation due to nonlinearity is a major problem for wideband SDR receivers. Any nonlinearity in the front end LNA can cause a strong

---

[51] Fundamentally, no component is completely linear—if nothing else, putting sufficient power in will cause it to burn up which is a very nonlinear effect.

[52] Power efficiency also often translates into size and weight. e.g., a less efficient amplifier needs a larger heat sink than a more efficient one.

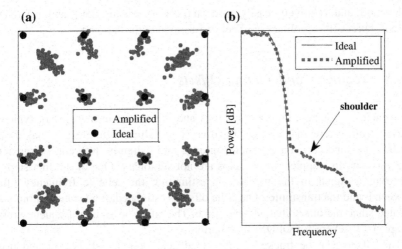

Fig. A.30 Effect of nonlinear amplification on pulse-shaped 16-QAM waveform **a** constellation, **b** spectrum

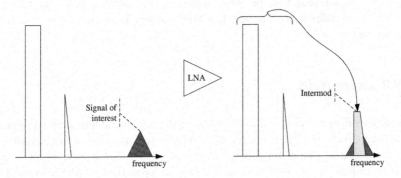

Fig. A.31 Intermodulation due to nonlinearity

intermodulation component to fall on top of the signal of interest (see Fig. A.31). Intermodulation is very difficult to mitigate with signal processing.[53]

A.9.7 DC Offsets and I/Q Imbalance

Zero if (direct conversion) RFFE suffer from constant offsets and relative gain mismatch on the I and Q branches (see Sect. 9.1.2). These effects are well

---

[53] It may be possible to examine the entire captured spectrum and estimate possible intermodulation products. MUD can then be potentially applied to remove the interference. To the best of the author's knowledge there is no published work on this topic.

understood and relatively easily mitigated as described [250] and references therein. Direct-IF sampling (see Sect. 9.1.3) avoids the problem.

A.9.8 Frequency and Timing Offsets

Frequencies for generating data symbols and the carrier in a radio are typically derived from a reference crystal. Each crystal is slightly different,[54] making the symbol rates and carrier frequencies in each radio slightly different. Frequencies also vary due to Doppler if the radios are not stationary. One of the first steps in acquiring a signal in the receiver is estimating the relative frequency offset between it and the transmitter. Once the offset is estimated, it must be continuously updated since the offset changes over time. The effect of residual frequency offset is especially severe for OFDM modulation.

The offsets can be tracked with a feedback loop or can be updated using feedforward estimator. The definitive reference to classical phase and timing synchronization techniques by Mengali [251] should be on every SDR designer's bookshelf. Synchronization becomes more difficult at low SNR. In fact, the powerful FEC codes operate at such low SNR that synchronization can become the limiting factor.

A.9.9 Phase Noise

Frequency offset and drift due to Doppler, crystal aging, and temperature variation change relatively slowly. However, the crystal also introduces a fast variation in the phase due to phase noise (also known as jitter). Phase noise is a complex process caused by a combination of physical properties of the crystal. It is characterized by the power spectral density of the phase variation [252, 253]. The power spectral density is specified in units of dBc/Hz—power relative to the tone at the nominal output frequency. Phase noise can be thought of as a random FM modulation. Instead of a single peak at the nominal frequency, the spectrum of a tone spreads out.

Most crystals have more phase noise power close to nominal frequency[55] as shown in Fig. A.32. The frequency of a crystal in a radio is scaled up to achieve the carrier frequency (and often the symbol frequency). Typical crystals oscillate

---

[54] Crystal accuracy is specified in parts per million (ppm). For example, a relatively good crystal has an absolute accuracy of 10 ppm. That means that a symbol rate of 10 MHz is accurate to ±100 Hz. A carrier frequency of 2 GHz is accurate to ±20 kHz. Frequency also changes over time and with temperature. More expensive (and typically larger and higher power) crystals have tighter specifications.

[55] One notable exception is rubidium atomic clocks which have an essentially constant phase noise power over frequency.

Fig. A.32 Typical phase
noise profile

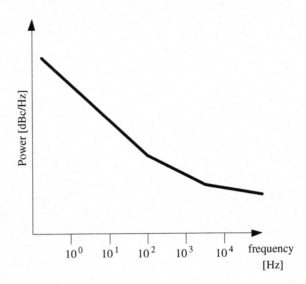

at 1–30 MHz. The phase noise power is increased by the scaling factor.

For example, a 10 MHz crystal has phase noise spec of -50 dBc/Hz at 1 kHz offset. The frequency is multiplied by 100 to get a carrier of 1 GHz. The phase noise at the carrier is then $-50 + 10 \log_{10} 100 = -30$ dBc/Hz.

Thus, phase noise becomes more of a concern at very high carrier frequencies (e.g. 60 GHz).

The effect of phase noise (just like any other frequency variation) depends on how frequently the phase estimate is updated in the receiver (usually once per symbol). Phase noise is less of a concern at high symbol rates because the phase is updated often. Since most of the phase noise power is concentrated at lower frequencies, the updates can keep up with the variations in phase and 'track them out.' However, at lower data rates, the phase can change by a large amount between two updates.

Appendix B
Recommended Test Equipment

Developing a SDR from scratch is a major undertaking. There is a very large gap between development of effective baseband DSP algorithms and a working radio suitable for field use. Most readers will have access to appropriate test equipment through their employers or schools. This section highlights some of the essential tools of the trade. All modern test equipment is designed to work with a PC for both control and data transfer. Ethernet interface is preferred to the older GPIB or serial interfaces.

- Digital oscilloscope (2+ channel, BW >500 MHz). Must be able to capture at least 10k samples. Desirable features include:

  - Digital inputs (8 or more)

- Spectrum analyzer (>3 GHz). Desirable features include:

  - Digital or analog IF output. The SA can be used as a downconverter.

- Vector signal generator (>3 GHz) with I/Q inputs. Desirable features include:

  - Integrated baseband signal generator (otherwise an arbitrary waveform generator is required).
  - Integrated noise generator.

- Two or more SDR development platforms. While it is possible to use a single board to both transmit and receive (loopback), two systems are required to verify tracking and acquisition.

  - High-end workstation with a programmable RFFE (e.g. USRP) or
  - FPGA-based platform.

- Channel emulator. A channel emulator should be specific for the intended SDR market and capabilities—a MIMO emulator for cellular communications is very different from a satellite channel emulator.

E. Grayver, *Implementing Software Defined Radio*,
DOI: 10.1007/978-1-4419-9332-8, © Springer Science+Business Media New York 2013

Optional:

- Arbitrary waveform generator (BW > 100 MHz).
- Noise generator
- Logic analyzer. This previously essential tool for debugging digital boards is becoming less useful with the advent of internal logic analyzers in FPGAs. An inexpensive logic analyzer can now be quickly rigged up using a cheap FPGA development board.

Appendix C
Sample XML files for an SCA Radio

The next few paragraphs discuss the contents of XML files used to define an SCA radio. For this example, a very simple SCA application, shown in Fig. C.33, was developed in OSSIE.[56] (Table C.1)

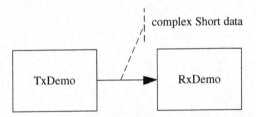

Fig. C.33 SCA application

---

[56] Unique identifiers (UIDs) are shortened for compactness

E. Grayver, *Implementing Software Defined Radio*,
DOI: 10.1007/978-1-4419-9332-8, © Springer Science+Business Media New York 2013

Table C.1 Software assembly descriptor (SAD) XML file—connecting the components

The SAD file pulls in 2 components, each defined in a component file. Components are defined in a Software Package Descriptor files

<pre><softwareassembly id="DCE:79aa" name="OSSIE::lab1">
  <componentfiles>
    <componentfile id="TxDemo_7e80" type="SPD">
      <localfile name="/xml/TxDemo/TxDemo.spd.xml"/>
    </componentfile>
    <componentfile id="RxDemo_7d0e" type="SPD">
      <localfile name="/xml/RxDemo/RxDemo.spd.xml"/>
    </componentfile>
  </componentfiles></pre>

DBRA OE must locate the components in order to work with them. In this example, the OE is told to locate the components by name, using the Naming Service.

<pre><partitioning>
  <componentplacement>
    <componentfileref refid="TxDemo_7e80"/>
    <componentinstantiation id="DCE:7e7b">
      <usagename>TxDemo_1</usagename>
      <findcomponent>
        <namingservice name="TxDemo_1"/>
      </findcomponent>
    </componentinstantiation>
  </componentplacement>
  <componentplacement>
    <componentfileref refid="RxDemo_7d0e"/>
    <componentinstantiation id="DCE:7d05">
      <usagename>RxDemo_1</usagename>
      <findcomponent>
        <namingservice name="RxDemo_1"/>
      </findcomponent>
    </componentinstantiation>
  </componentplacement>
</partitioning></pre>

Assembly controller specifies which component is responsible for starting and stopping the signal chain. In this case, everything starts from the TxDemo

<pre><assemblycontroller>
  <componentinstantiationref refid="DCE:7e7b"/>
</assemblycontroller></pre>

Connections section specifies the ports on each component and how the ports are connected. Note that in SCA terminology a "provides" port is an input and a "uses" port is an output. Thus, RxDemo has an input port symbols_in, while TxDemo has an output port symbols_out. Each point-to-point connection (uses→provides) is defined in a separate `<connectinterface>` (only one in this example)

```
<connections>
  <connectinterface id="DCE:79b1">
    <providesport>
      <providesidentifier>symbols_in</providesidentifier>
      <findby>
        <namingservice name="RxDemo_1"/>
      </findby>
    </providesport>
    <usesport>
      <usesidentifier>symbols_out</usesidentifier>
      <findby>
        <namingservice name="TxDemo_1"/>
      </findby>
    </usesport>
  </connectinterface>
</connections>
```

Let us now examine what the SCA OE knows about the TxDemo component. The component is referenced from the SAD file with a link to the **TxDemo.spd.xml** file, shown in Table C.2. (Tables C.3, C.4, C.5)

Table C.2 Software package descriptor (SPD) XML file—setting the executable filename

```
<softpkg id="DCE:23ec" name="TxDemo">
  <title/>
  <description>QPSK symbols</description>
  <author>
    <name>OSSIE Project</name>
    <company>MPRR Group</company>
    <webpage>http://www.mprg.org</webpage>
  </author>
```

Name will be used to find this component. Information about the author and description is only used for reference and by GUI tools.

```
<propertyfile type="PRF">
  <localfile name="/xml/TxDemo/TxDemo.prf.xml"/>
</propertyfile>
```

A property file describes run-time configurable parameters for the component (e.g. *gain* for an amplifier). See Table C-3

```
<descriptor>
  <localfile name="/xml/TxDemo/TxDemo.scd.xml"/>
</descriptor>
```

The descriptor (SCD) file contains information about input and output ports.

```
<implementation id="DCE:23ed">
  <description>Linux X86 implementation</description>
  <code type="Executable">
    <localfile name="/bin/TxDemo"/>
  </code>
  <os name="Linux" version="2.6.26.3"/>
  <processor name="x86"/>
</implementation>
<implementation id="DCE:aa6f">
  <description>Linux PPC implementation</description>
  <code type="Executable">
    <localfile name="/bin/TxDemo_powerpc"/>
  </code>
  <os name="Linux" version="2.6.27-rc9-xlnx"/>
  <processor name="powerpc"/>
</implementation>
</softpkg>
```

The same component may be available in multiple implementations. One version may run on an x86 processor, another on a PowerPC, while yet another requires an FPGA. Each implementation will be examined by the OE until a match is found between the implementation requirements (e.g. processor type) and available (or assigned) resources. The key element for each implementation is the localfile, which tells OE which binary to execute. The TxDemo component is available for x86 and PowerPC processors.

Table C.3 Properties descriptor (PRF) XML file—defining parameters

```
<properties>
  <simple id="DCE:1b9c" mode="readonly"
   name="packet_delay_ms" type="short">
    <value>1000</value>
    <description>packet delay</description>
    <kind kindtype="configure"/>
  </simple>
</properties>
```

The TxDemo component has one configurable parameter – delay between packets. The parameter is a 16-bit integer. The parameter is read-only, meaning that it cannot be changed at runtime. Properties that are not read-only can be set by DBRA *configure()* function. Properties that are not write-only can be checked by *query()* function.

Table C.4 Software component descriptor (SCD) XML file—defining the IO ports

```<softwarecomponent>```
  ```<corbaversion>2.2</corbaversion>```
  ```<componentrepid repid="IDL:CF/Resource:1.0"/>```
  ```<componenttype>resource</componenttype>```

SCD files are used to describe almost every object in the SCA pantheon. All the signal processing components are of the type *resource*.

```<componentfeatures>```
  ```<supportsinterface repid="IDL:CF/Resource:1.0"```
   ```supportsname="Resource"/>```
  ```<supportsinterface repid="IDL:CF/LifeCycle:1.0"```
   ```supportsname="LifeCycle"/>```
  ```<supportsinterface repid="IDL:CF/PortSupplier:1.0"```
   ```supportsname="PortSupplier"/>```
  ```<supportsinterface repid="IDL:CF/PropertySet:1.0"```
   ```supportsname="PropertySet"/>```
  ```<supportsinterface repid="IDL:CF/TestableObject:1.0"```
   ```supportsname="TestableObject"/>```

A number of standard interfaces are defined for SCA components. This section explicitly states which standard interfaces are supported by this component. For example, the TxDemo module has *ports*, and *properties*.

```<ports>```
  ```<uses repid="IDL:standardInterfaces/complexShort:1.0"```
   ```usesname="symbols_out">```
   ```<porttype type="data"/>```
  ```</uses>```
```</ports>```

The component input and output ports are defined here. These ports will be referenced in the SAD file to connect the components. Note the **symbols_out** port, which was referenced on row 7 of Table C-1. The data type (complex 16-bit values) is referenced from the set of standard SCA data types.

```<interfaces>```
  ```<interface name="Resource" repid="IDL:CF/Resource:1.0">```
   ```<inheritsinterface repid="IDL:CF/LifeCycle:1.0"/>```
   ```<inheritsinterface repid="IDL:CF/PortSupplier:1.0"/>```
   ```<inheritsinterface repid="IDL:CF/PropertySet:1.0"/>```
   ```<inheritsinterface repid="IDL:CF/TestableObject:1.0"/>```
  ```</interface>```
  ```<interface name="complexShort"```
      ```repid="IDL:standardInterfaces/complexShort:1.0"/>```
```</interfaces>```

The interfaces mirror (typically) the component features section in row 3.

Table C.5 Deployment enforcement XML file—defining devices to execute components

```
<deploymentenforcement>
  <application id="DCE:438bf " name="Name"/>
  <deviceassignmentsequence>
    <deviceassignmenttype>
      <componentid>DCE:7e7b</componentid>
      <assigndeviceid>DCE:b461</assigndeviceid>
    </deviceassignmenttype>
    <deviceassignmenttype>
      <componentid>DCE:7d05</componentid>
      <assigndeviceid>DCE:5ba3</assigndeviceid>
    </deviceassignmenttype>
  </deviceassignmentsequence>
</deploymentenforcement>
```

SCA does not specify a way force execution of a particular on a specific device [252]. Some SCA OE providers developed a non-standard way to configure this 'deployment.' This table shows an example of such a file for OSSIE. Note that components TxDemo and RxDemo are configured to execute on different devices (GPP and FPGA in this case).

Bibliography

1. Proakis M (2007) Digital communications. McGraw-Hill, New York
2. Sklar B (2001) Digital communications: fundamentals and applications. Prentice-Hall PTR, Englewood Cliffs
3. Wireless Innovation Forum (2011) Driving the future of radio communications and systems worldwide. http://www.wirelessinnovation.org
4. Zimmermann H (1980) OSI reference model–the ISO model of architecture for open systems interconnection. Commun, IEEE Trans 28:425–432
5. Hanzo L, Ng SX, Webb WT, Keller T (2004) Quadrature amplitude modulation: from basics to adaptive trellis-coded, turbo-equalised and space-time coded OFDM, CDMA and MC-CDMA systems. IEEE Press-John Wiley, New York
6. Caire G, Kumar KR (2007) Information theoretic foundations of adaptive coded modulation. Proc IEEE 95:2274–2298
7. James JH, Milton S, Bomey Y, Zhong Y (2008) Using FEC, channel Interleaving, ARQ, and DCM to mitigate fades of various timescales. In: 26th international communications satellite systems conference (ICSSC)
8. Smolnikar M, Javornik T, Mohorcic M, Berioli M (2008) DVB-S2 adaptive coding and modulation for HAP communication system. In: Vehicular technology conference, VTC spring 2008, IEEE. pp 2947–2951
9. Sbarounis C, Squires R, Smigla T, Faris F, Agarwal A (2004) Dynamic bandwidth and resource allocation (DBRA) for MILSATCOM. In: Military communications conference, 2004. MILCOM 2004, IEEE, vol 2. pp 758–764
10. Kivanc D, Guoqing L, Hui L (2003) Computationally efficient bandwidth allocation and power control for OFDMA. Wirel Commun, IEEE Trans, vol 2. pp 1150–1158
11. Koffman I, Roman V (2002) Broadband wireless access solutions based on OFDM access in IEEE. Commun Mag, IEEE 40:96–103
12. Rhee W, Cioffi JM (2000) Increase in capacity of multiuser OFDM system using dynamic. In: Vehicular technology conference proceedings, VTC 2000-Spring Tokyo. 2000 IEEE 51st, vol 2. pp 1085–1089
13. Mitola J, Maguire GQ (1999) Cognitive radio: making software radios more personal. Pers Commun, IEEE 6:13–18
14. Haykin S (2005) Cognitive radio: brain-empowered wireless communications. Sel Areas Commun, IEEE J 23:201–220
15. National Telecommunications and Information Admini (2003) United States frequency allocations: the radio spectrum. http://www.ntia.doc.gov/osmhome/allochrt.pdf
16. Doumi TL (2006) Spectrum considerations for public safety in the United States. Commun Mag, IEEE 44:30–37

E. Grayver, *Implementing Software Defined Radio*,
DOI: 10.1007/978-1-4419-9332-8, © Springer Science+Business Media New York 2013

17. Yucek T, Arslan H (2009) A survey of spectrum sensing algorithms for cognitive radio applications. Commun Surv Tutorials, IEEE 11:116–130
18. Le B et al (2007) A public safety cognitive radio node. In: SDR Forum, Denver
19. He A et al (2010) A survey of artificial intelligence for cognitive radios. Veh Technol, IEEE Trans 59:1578–1592
20. Mary P, Gorce JM, Villemaud G, Dohler M, Arndt M (2007) Reduced complexity MUD-MLSE receiver for partially-overlapping WLAN-like interference. In: Vehicular technology conference, 2007. VTC2007-Spring. IEEE 65th, pp 1876–1880
21. Congzheng Han et al (2011) Green radio: radio techniques to enable energy-efficient wireless networks. Commun Mag, IEEE 49:46–54
22. Gür G, Alagöz F (2011) Green wireless communications via cognitive dimension: an overview. Network, IEEE 25:50–56
23. Chen Y, Zhang S, Xu S, Li GY (2011) Fundamental trade-offs on green wireless networks. Commun Mag, IEEE 49:30–37
24. Grace D, Chen J, Jiang T, Mitchell PD (2009) Using cognitive radio to deliver 'Green' communications. In: Cognitive radio oriented wireless networks and communications, CROWNCOM '09. 4th international conference on 2009. pp 1–6
25. Viswanath P, Tse DN, Laroia R (2002) Opportunistic beamforming using dumb antennas. Inf Theor, IEEE Trans 48:1277–1294
26. Bogucka H, Conti A (2011) Degrees of freedom for energy savings in practical adaptive wireless systems. Commun Mag, IEEE 49:38–45
27. Popken L (2007) Rescuing the huygens mission from fiasco. Proc IEEE 95:2248–2258
28. Deutsch LJ (2002) Resolving the cassini/huygens relay radio anomaly. In: IEEE aerospace conference, Big Sky, MT, pp 3-1295–3-1302. http://trs-new.jpl.nasa.gov/dspace/bitstream/2014/13159/1/01-1802.pdf
29. Vallés EL, Prabhu RS, Dafesh PA (2010) Phase tracking with a cognitive antijam receiver system (CARS). In: Military communications conference, MILCOM 2010. pp 1701–1706
30. Prabhu RS, Valles EL, Dafesh PA (2009) Cognitive anti-jam radio system. Software Defined Radio Forum, Washington
31. Grayver E et al (2009) Cross-layer mitigation techniques for channel impairments. In: Aerospace conference, 2009 IEEE. pp 1–9
32. Ixia (2011) IxN2X multiservice test solution. www.ixiacom.com/products/ixn2x/index.php
33. Grayver E, Chen J, Utter A (2008) Application-layer codec adaptation for dynamic bandwidth resource allocation. In: Aerospace conference, 2008 IEEE. pp 1–8
34. Hant J, Swaminathan V, Kim J (2008) Real-time emulation of internet applications over satellite links using the SAtellite link EMulator (SALEM). In: AIAA, San Deigo
35. Murmann B (2008) A/D converter trends: power dissipation, scaling and digitally assisted architectures. In: Custom integrated circuits conference, 2008. CICC 2008. IEEE. pp 105–112
36. Michael LB, Mihaljevic MJ, Haruyama S, Kohno R (2002) A framework for secure download for software-defined radio. Commun Mag, IEEE 40:88–96
37. Dong X (2006) Effect of slow fading and adaptive modulation on TCP/UDP performance of high-speed packet wireless networks. University of California at Berkeley, Berkeley, Technical Report UCB/EECS-2006-109
38. Kuon I, Rose J (2006) Measuring the gap between FPGAs and ASICs. pp 21–30
39. FPGA Developer (2011) List and comparison of FPGA companies. http://www.fpgadeveloper.com/2011/07/list-and-comparison-of-fpga-companies.html
40. Rezgui S, Wang JJ, Sun Y, Cronquist B, McCollum J (2008) New reprogrammable and non-volatile radiation tolerant FPGA: RTA3P. In: Aerospace conference, 2008 IEEE. pp 1–11
41. Microsemi (2011) RT proASIC3 FPGAs. http://www.actel.com/products/milaero/rtpa3/default.aspx
42. Microsemi (2011) Axcelerator antifuse FPGAs. http://www.actel.com/products/axcelerator/default.aspx

43. Palkovic M et al (2010) Future software-defined radio platforms and mapping flows. Sig Process Mag, IEEE 27:22–33
44. Farber R (2011) CUDA application design and development. Morgan Kaufmann
45. KHRONOS Group (2011) OpenCL—the open standard for parallel programming of heterogeneous systems. http://www.khronos.org/opencl/
46. Munshi A, Gaster B, Mattson TG, Fung J, Ginsburg D (2011) OpenCL programming guide. Addison-Wesley Professional, Boston
47. Accelereyes (2011) Jacket GPU engine. http://www.accelereyes.com/products
48. Kahle JA et al (2005) Introduction to the cell multiprocessor
49. Lin Y et al (2007) SODA: a high-performance DSP architecture for software-defined radio. Micro, IEEE 27:114–123
50. Woh M et al (2008) From SODA to scotch: the evolution of a wireless baseband processor. In: Microarchitecture, 41st IEEE/ACM international symposium on, pp 152–163
51. Williston K (2009) Massively parallel processors: who's still alive?. http://www.eetimes.com/electronics-blogs/dsp-designline-blog/4034360/Massively-parallel-processors-Who-s-still-alive
52. Schulte Michael et al (2006) A low-power multithreaded processor for software defined radio. J VLSI Sig Process Syst 43(2–3):143–159
53. Mamidi S et al (2005) Instruction set extensions for software defined radio on a multithreaded processor. pp 266–273
54. Coherent Logix (2011) HyperX. http://www.coherentlogix.com/
55. Octasic. OCT2224W is a system-on-chip. http://www.octasic.com/en/products/oct2200/oct2224w.php
56. picoChip. PC205 high-performance signal processor. http://www.picochip.com/page/76/
57. Tilera. Multicore Processing. http://tilera.com/
58. Malone M (2008) On-board processing expandable reconfigurable architecture (OPERA) program overview. In: The first workshop on fault-tolerant spaceborne computing employing new technologies. Albuquerque, http://www.zettaflops.org/spc08/Malone-Govt-OPERA-FT-Spaceborne-Computing-Workshop-rev3.pdf
59. Singh K et al (2011) FFTW and complex ambiguity function performance on the maestro processor. In: Aerospace conference, 2011 IEEE. pp 1–8
60. Walters JP, Kost R, Singh K, Suh J, Crago SP (2011) Software-based fault tolerance for the maestro many-core processor. In: Aerospace conference, 2011 IEEE. pp 1–12
61. Texas Instruments (2009) OMAP^{TM} 1 processors. http://focus.ti.com/general/docs/wtbu/wtbuproductcontent.tsp?templateId=6123&navigationId=11991&contentId=4670
62. Xilinx (2011) Zynq-7000 extensible processing platform. http://www.xilinx.com/products/silicon-devices/epp/zynq-7000/index.htm
63. Berkeley Design Technology (2006) Enabling technologies for SDR: comparing FPGA and DSP performance. http://www.bdti.com/MyBDTI/pubs/20061115_sdr06_fpgas.pdf
64. Valles EL, Tarasov K, Roberson J, Grayver E, King K (2009) An EMWIN and LRIT software receiver using GNU radio. In: Aerospace conference, 2009 IEEE. pp 1–11
65. Feng W, Balaji P, Baron C, Bhuyan LN, Panda DK (2005) Performance characterization of a 10-gigabit ethernet TOE, high-performance interconnects, symposium on, vol 0, pp 58–63
66. Hughes-Jones R, Clarke P, Dallison S (2005) Performance of 1 and 10 Gigabit ethernet cards with server quality motherboards. Future Gener Comput Syst 21:469–488
67. Schmid T, Sekkat O, Srivastava MB (2007) An experimental study of network performance impact of increased latency in software defined radios. pp 59–66
68. Mills D (2007) Network time protocol version 4 protocol and algorithm specification. IETF, RFC 1305
69. Eidson JC, Fischer M, White J (2002) IEEE-1588 standard for a precision clock synchronization protocol for networked measurement and control systems. IEEE, Standard 1588
70. Dobkin DM (2007) The RF in RFID: passive UHF RFID in practice. Elsevier/Newnes, p 34

71. Lu G, Lu P (2008) A software defined radio receiver architecture based on cluster computing. In: Grid and cooperative computing, GCC '08, Seventh international conference on 2008. pp 612–619

72. Papadimitriou K, Dollas A (2006) Performance of partial reconfiguration in FPGA systems: a survey and a cost model. http://ee.washington.edu/faculty/hauck/publications/KypPart Reconfig.pdf

73. Yuan M, Gu Z, He X, Liu X, Jiang L (2010) Hardware/software partitioning and pipelined scheduling on runtime reconfigurable FPGAs. ACM Trans Des Autom Electron Syst 15(2):13:1–13:41

74. McDonald EJ (2008) Runtime FPGA partial reconfiguration. In: IEEE aerospace conference, Big Sky

75. Suris J et al (2008) Untethered on-the-fly radio assembly with wires-on-demand. In: Aerospace and electronics conference, NAECON 2008, IEEE National. pp 229–233

76. JPEO JTRS (2011) SCA: application program interfaces (APIs). http://sca.jpeojtrs.mil/api.asp

77. Xilinx (2011) Xilinx platform flash XL. http://www.xilinx.com/products/config_mem/pfxl.htm

78. Altera (2011) Altera parallel flash loader megafunction. http://www.altera.com/literature/ug/ug_pfl.pdf

79. Parhi K (1999) VLSI digital signal processing systems: design and implementation. Wiley, New York

80. Hussein J, Klein M, Hart M (2011) Lowering power at 28 nm with Xilinx 7 series FPGAs. Xilinx, white paper WP389

81. Altera (2010) Reducing power consumption and increasing bandwidth on 28 nm FPGAs. White paper WP-01148-1.0

82. Edwards M (1997) Software acceleration using coprocessors: is it worth the effort?. In: Hardware/software codesign, (CODES/CASHE '97), Proceedings of the 5th international workshop on 1997. pp 135–139

83. Lange H (2011) Reconfigurable computing platforms and target system architectures for automatic HW/SW compilation. TUt Darmstadt, Darmstadt, PhD thesis, http://tuprints.ulb.tu-darmstadt.de/2560/

84. Intel (2011) VTune™ performance analyzer. http://software.intel.com/en-us/articles/intel-vtune-amplifier-xe/

85. Tian X, Benkrid K (2010) High-performance quasi-monte carlo financial simulation: FPGA vs. GPP vs. GPU. ACM Trans Reconfigurable Technol Syst 3(4):26:1–26:22

86. Nallatech PCIe-280–8-lane PCI express 2.0 accelerator. http://www.nallatech.com/Motherboards/pcie-280-8-lane-pci-express-20-accelerator-card-featuring-xilinx-virtex-5-fpga-and-memory.html

87. Nvidia (2011) TESLA C2050 and TESLA C2070 computing processor board. http://www.nvidia.com/docs/IO/43395/BD-04983-001_v04.pdf

88. eVGA (2010) EVGA classified SR-2 motherboard. http://www.evga.com/articles/00537/

89. Magma (2011) ExpressBox7 (x8) Gen2. http://magma.com/expressbox7x8G2.asp

90. Nallatech (2011) Intel xeon FSB FPGA socket fillers. http://www.nallatech.com/intel-xeon-fsb-fpga-socket-fillers.html

91. XtremeData (2009) XD2000F™ FPGA in-socket accelerator for AMD socket F. http://old.xtremedatainc.com/index.php?option=com_content&view=article&id=105&Itemid=59

92. Reinhart RC et al (2007) Open architecture standard for NASA's software-defined space telecommunications radio systems. Proc IEEE 95:1986–1993

93. JPEO JTRS http://sca.jpeojtrs.mil/home.asp

94. Bard J (2007) Software defined radio: the software communications architecture. Wiley, New York

95. Ciaran McHale, CORBA explained simply. http://www.ciaranmchale.com/corba-explained-simply/

96. Balister PJ, Robert M, Reed JH (2006) Impact of the use of CORBA for inter-component communication in SCA based radio. In: SDR forum technical conference, Washington

97. Ulversøy T, Neset JO (2008) On workload in an SCA-based system, with varying component and data packet sizes. NATO, RTO-MP-IST-083

98. Cormier AR, Dietrich CB, Price J, Reed JH (2010) Dynamic reconfiguration of software defined radios using standard architectures. Phys Commun 3:73–80

99. Stephens DR, Magsombol C, Jimenez C (2007) Design patterns of the JTRS infrastructure. In: Military communications conference, MILCOM 2007, IEEE. pp 1–5

100. Magsombol C, Jimenez C, Stephens DR (2007) Joint tactical radio system—application programming interfaces. In: Military communications conference, MILCOM 2007, IEEE. pp 1–7

101. ISO/IEC (2002) High-level data link control (HDLC) procedures. http://webstore.iec.ch/preview/info_isoiec13239%7Bed3.0%7Den.pdf

102. National Geospatial-Intelligence Agency (2009) Grids and reference systems. http://earth-info.nga.mil/GandG/coordsys/grids/referencesys.html

103. Reinhart RC et al (2010) Space telecommunications radio system (STRS) architecture standard. NASA glenn research center, Clevelend, TM 2010-216809

104. Willink E (2002) The waveform description language. Wiley, New York

105. Willink ED (2001) The waveform description language: moving from implementation to specification. In: Military communications conference, MILCOM 2001, communications for network-centric operations: creating the information force, IEEE, vol 1. pp 208–212

106. Scaperoth D et al (2006) Cognitive radio platform development for interoperability. In: Military communications conference, MILCOM 2006, IEEE. pp 1–6

107. Grayver E (2009) Physical layer description standard for software defined radio. Software Defined Radio Forum, Washington

108. Reed Solomon codes as BCH codes http://en.wikipedia.org/wiki/Reed-Solomon_code#Reed-Solomon_codes_as_BCH_codes

109. Comtech, CDM-750 advanced high-speed trunking modem. http://www.comtechefdata.com/products/modems/pCDM-750.asp

110. Pucker L (2005) Does the wireless industry really need all these digital if standards? Commun Mag, IEEE 43:54–57

111. VITA, VITA Radio Transport (VRT) Standard. https://www.vita.com/online-store.html

112. Normoyle R, Mesibov P (2008) The VITA radio transport as a framework for software definable radio architectures. In: Software Defined Radio Forum

113. Vigil AJ (2010) Digital IF for MILSATCOM terminals: CONOPS and standardization for the next decade and beyond. In: Military communications conference, MILCOM 2010. pp 1128–1133

114. thinkRF (2011) WSA4000 RF digitizer. http://www.thinkrf.com/wsa4000_spec.html

115. Beljour H et al (2010) Proof of concept effort for demonstrating an all-digital satellite communications earth terminal. In: Military communications conference, MILCOM 2010. pp 1547–1551

116. MIPI Alliance. DigRF(SM) Specifications http://www.mipi.org/specifications/digrfsm-specifications

117. Seng E (2009) Testing digRF for 3G handsets. Test and Measurement World, May

118. Innovative Computer Engineering (2011) SDDS packet format. www.ice-online.com/BACKUP/SDDSPacket.doc

119. OBSAI (2010) RP3 specification. http://www.obsai.org/obsai/documents/public_documents/download_specifications/rp_specifications

120. Lanzani CF (2007) OBSAI RP3-01 6.144 Gbps interface implementation. FPGAworld

121. CPRI (2011) Common public radio interface specification. http://www.cpri.info/spec.html

122. Varghese AA, Linder LF (2008) Software defined radio—the "Holy Grail" for the industry. Wirel Des Mag, April 2008

123. Intel (2011) Integrated performance primitives (IPP). http://software.intel.com/en-us/articles/intel-ipp/
124. Lu Z (2010) How GNU radio core works—an analysis of GNU radio 3.3.0 sources. http://apachepine.ece.ncsu.edu/resources/How%20GNU%20Radio%20Core%20Works.pdf
125. The GOES-R program office, GOES-R main website. http://www.goes-r.gov/
126. National Oceanic and Atmospheric Administration (NOAA) EMWIN-N overview. http://www.weather.gov/emwin/index.htm
127. National Oceanic and Atmospheric Administration (NOAA) LRIT Overview. http://noaasis.noaa.gov/LRIT
128. Mueller K, Muller M (1976) Timing recovery in digital synchronous data receivers. Commun, IEEE Trans 24:516–531
129. Palomo A, Villing R, Farrell R (2008) Software defined radio architectures evaluation. In: SDR 08 technical conference and product exposition, Washington
130. Chen D, Vanhoy G, Beaufait M, Dietrich CB (2010) OSSIE/GNU radio generic component demonstration. In: SDR '10 technical conference, Washington
131. Prismtech, Spectra CX software communications architecture (SCA) tool. http://www.prismtech.com/spectra/products/spectra-cx
132. CRC, SCARI software suite. http://www.crc.gc.ca/en/html/crc/home/research/satcom/rars/sdr/products/scari_suite/scari_suite
133. Tan K et al (2011) Sora: high-performance software radio using general-purpose multi-core processors. Commun ACM 54(1):99–107
134. Falcao G, Sousa L, Silva V (2011) Massively LDPC decoding on multicore architectures. Parallel Distrib Syst, IEEE Trans 22:309–322
135. Hwang J (2003) Innovative communication design lab based on PC sound card and matlab: a software-defined-radio OFDM modem example. In: Acoustics, speech, and signal processing, Proceedings (ICASSP '03), IEEE international conference on 2003, vol 3. pp III–761-4
136. Beaudoin L et al (2003) Receiving images directly from meteorological satellites in an engineers' school: technical and pedagogical aspects. In: Geoscience and remote sensing symposium, 2003, IGARSS '03, Proceedings, IEEE international, vol 1. pp 479–481
137. Avtec Systems, Inc (2005) EMWIN OQPSK transmit and receiver prototype specifications. NOAA/CSC, TR 05-004
138. Elektor (2011) Software defined radio. http://www.elektor.com/products/kits-modules/modules/software-defined-radio-%28070039-91%29.91475.lynkx
139. Mathworks (2011) MATLAB and Simulink support for USRP^{TM} devices. http://www.mathworks.com/discovery/sdr/usrp.html
140. National Instruments (2011) What Is NI USRP hardware? http://zone.ni.com/devzone/cda/tut/p/id/12985
141. Hamed Firooz (2011) Implementation of full-bandwidth 802.11b receiver. http://span.ece.utah.edu/pmwiki/pmwiki.php?n=Main.80211bReceiver
142. Sparkfun (2011) SiGe GN3S sampler v2. http://www.sparkfun.com/products/8238
143. Borre K, Akos DM (2006) A software-defined GPS and galileo receiver: a single-frequency approach (applied and numerical harmonic analysis). Birkhäuser, Boston
144. NSL (2011) GNSS SDR front end and receiver. http://www.nsl.eu.com/primo.html
145. Rincon, DLSR-1U-xx DSPBrik L-band software radio. http://www.rincon.com/DLSR_1U_xx_Final.pdf
146. Midwest Microwave Solutions (2011) VHF/UHF digitizers. http://mms-rf.com/rf-digitizers.html
147. Innovative Computer Engineering, GIGExD. http://www.ice-online.com/rumeL/RI-GIGEXD.htm
148. EPCOS (2007) Front-end solutions for world phones. http://www.epcos.com/web/generator/Web/Sections/Components/Page,locale=en,r=263282,a=575612.html

149. Ghannouchi FM (2010) Power amplifier and transmitter architectures for software defined radio systems. Circ Syst Mag, IEEE 10:56–63
150. Hentschel T (2005) The six-port as a communications receiver. Microwave Theor Tech, IEEE Trans 53:1039–1047
151. Xu X, Wu K, Bosisio RG (2003) Software defined radio receiver based on six-port technology. In: Microwave symposium digest, 2003 IEEE MTT-S international, vol 2. pp 1059–1062
152. Agilent Inc (2011) See the future of arbitrary waveform generators. http://www.home.agilent.com/upload/cmc_upload/All/March-1-2011-AWG-webcast.pdf
153. Hunter MT, Kourtellis AG, Ziomek CD, Mikhael WB (2010) Fundamentals of modern spectral analysis. In: AUTOTESTCON, IEEE, pp 1–5
154. Ingels M et al (2010) A 5 mm², 40 nm LP CMOS transceiver for a software-defined radio platform. Solid-State Circuits, IEEE J 45:2794–2806
155. Giannini V (2008) Baseband analog circuits for software defined radio. Springer
156. Lime Microsystems (2011) Multi-band multi-standard transceiver with digital interface (LMS6002D). http://www.limemicro.com/download/LMS6002D_Product_Brief.pdf
157. Walden RH (2008) Analog-to-digital conversion in the early 21st century. In: Benjamin W (ed) Wiley encyclopedia of computer science and engineering. Wiley, Hoboken
158. Murmann B (2011) ADC performance survey. http://www.stanford.edu/~murmann/adcsurvey.html
159. DARPA Microsystems Technology Office (2009) Remoted analog-to-digital converter with de-serialization and reconstruction. DARPA, BAA 09-51
160. Poberezhskiy YS, Poberezhskiy GY (2004) Sampling and signal reconstruction circuits performing internal antialiasing filtering and their influence on the design of digital receivers and transmitters. Circuits Syst I: Regul Pap, IEEE Trans 51:118–129
161. Schreier R (2005) Understanding delta-sigma data converters. IEEE Press, New York
162. Murmann B, Daigle C (2010) DAC survey [private communications]
163. Shoop B (2001) Photonic analog-to-digital conversion. Springer, Berlin
164. Bussjager R (2006) Photonic analog-to-digital converters. Air Force Research Lab, Rome, NY, AFRL-SN-RS-TR-2006-109
165. Grein ME et al (2011) Demonstration of a 10 GHz CMOS-compatible integrated photonic analog-to-digital converter. In: CLEO:2011—laser applications to photonic applications, p CThI1
166. Yao J (2009) Microwave photonics. Lightwave Technol, J 27:314–335
167. Cox III CH, Ackerman EI (2011) Photonics for simultaneous transmit and receive. In: International microwave symposium, Baltimore
168. Mukhanov OA, Gupta D, Kadin AM, Semenov VK (2004) Superconductor analog-to-digital converters. Proc IEEE 92:1564–1584
169. HYPRES, Inc (2011) Digital-RF receiver. http://www.hypres.com/products/digital-rf-receiver/
170. Gorju G et al (2007) 10 GHz Bandwidth rf spectral analyzer with megahertz resolution based on spectral-spatial holography in Tm3$+$:YAG: experimental and theoretical study. J Opt Soc Am B 24:457–470
171. Kunkee E et al (2008) Photonically enabled RF spectrum analyzer demonstration
172. Colice M, Schlottau F, Wagner KH (2006) Broadband radio-frequency spectrum analysis in spectral-hole-burning media. Appl Opt 45:6393–6408
173. Llamas-Garro I, Brito-Brito Z (2010) Reconfigurable microwave filters. In: Minin I (ed) Microwave and millimeter wave technologies: from photonic bandgap devices to antenna and applications. Chap. 7, pp 185–205
174. K&L Microwave (2011) Tunable filter products. http://www.klmicrowave.com
175. Roy MK, Richter J (2006) Tunable ferroelectric filters for software defined tactical radios. In: Applications of ferroelectrics, isaf '06, 15th IEEE international symposium on the 2006. pp 348–351

176. Agile RF (2011) http://www.agilerf.com/products/filters.html
177. Wang X, Bao P, Jackson TJ, Lancaster MJ (2011) Tunable microwave filters based on discrete ferroelectric and semiconductor varactors. Microwaves, Antennas Propag, IET 5:776–782
178. Kwang C, Courreges S, Zhiyong Z, Papapolymerou J, Hunt A (2009) X-band and Ka-band tunable devices using low-loss BST ferroelectric capacitors. In: Applications of ferroelectrics, ISAF 2009. 18th IEEE international symposium on the 2009, pp 1–6
179. Hittite Microwave Corporation (2011) Tunable bandpass filters. http://www.hittite.com/products/index.html/category/342
180. Fjerstad RL (1970) Some design considerations and realizations of iris-coupled YIG-tuned filters in the 12–40 GHz region. Microwave Theory Tech, IEEE Trans 18:205–212
181. OMNIYIG, INC (2011) http://www.omniyig.com/
182. Giga-tronics (2011) http://www.gigatronics.com/product/c/tunable-microwave-filters
183. Teledyne Microwave, YIG filter information. http://www.teledynemicrowave.com/pdf_YIG/Teledyne%20Microwave%20YIG%20Filter%20Products%200507.pdf
184. Entesari K (2006) Development of high performance 6–18 GHz tunable/switchable RF. The University of Michigan, Ann Arbor, PhD thesis
185. Alipour P et al (2011) Fully reconfigurable compact RF photonic filters using high-Q silicon microdisk resonators. In: Optical fiber communication conference and exposition (OFC/NFOEC), 2011 and the national fiber optic engineers conference, pp 1–2
186. Norberg EJ, Guzzon RS, Parker JS, Johansson LA, Coldren LA (2011) Programmable photonic microwave filters monolithically integrated in InP–InGaAsP. Lightwave Technol, J 29:1611–1619
187. Savchenkov AA et al (2009) RF photonic signal processing components: from high order tunable filters to high stability tunable oscillators. In: Radar conference 2009. IEEE, pp 1–6
188. Constantine Balanis (2005) Antenna theory: analysis and design. Wiley, New york
189. Cetiner BA et al (2004) Multifunctional reconfigurable MEMS integrated antennas for adaptive MIMO systems. Commun Mag, IEEE 42:62–70
190. Maddela M, Ramadoss R, Lempkowski R (2007) PCB MEMS-based tunable coplanar patch antenna. In: industrial electronics, ISIE 2007. IEEE international symposium on 2007, pp 3255–3260
191. Huang L, Russer P (2008) Electrically tunable antenna design procedure for mobile applications. Microwave Theory Tech, IEEE Trans 56:2789–2797
192. Lee EY et al (2009) The software defined antenna: measurement and simulation of a 2 element array. In: Antennas and propagation society international symposium, 2009. APSURSI '09. IEEE, pp 1–4
193. Kumar R (2011) Plasma antenna. Lambert Academic Publishing
194. Jenn DC (2003) Plasma antennas: survey of techniques and the current state of the art. Naval Posgraduate School, Monterrey, CA, technical report NPS-CRC-03-001
195. Plasma Antennas (2011) PSiAn plasma antennas. http://www.plasmaantennas.com
196. AARONIA AG (2011) Logarithmic periodic antenna. http://www.aaronia.com/Datasheets/Antennas/Broadband_Measurementantenna_HyperLOG3000.pdf
197. Zhao J, Chen C-C, Volakis JL (2010) Low profile ultra-wideband antennas for software defined radio. In: Antennas and propagation society international symposium (APSURSI), 2010 IEEE, pp 1–4
198. IBM, Rational DOORS. http://www.ibm.com/software/awdtools/doors/
199. OSRMT (2010) Open source requirements management tool. http://sourceforge.net/projects/osrmt/
200. Lamsweerde A (2009) Requirements engineering : from system goals to UML models to software specifications. Wiley, New York
201. Bailey B (2010) ESL models and their application electronic system level design and verification in practice. Springer, Berlin

202. Fingeroff M (2010) High-level synthesis: blue book. Xlibris Corporation Mentor Graphics Corporation, Philadelphia
203. Knapp D (1996) Behavioral synthesis: digital system design using the synopsys behavioral compiler. Prentice Hall PTR, Englewood Cliffs
204. Cong J et al (2011) High-level synthesis for FPGAs: from prototyping to deployment. Comput. Aided Des Integr Circ Syst, IEEE Trans 30:473–491
205. Morris K (2011) C to FPGA: who'll use the next generation of design tools? Electron Eng J, June 2011. http://www.ecjournal.com/archives/articles/20110621-nextgen/
206. Xilinx (2011) System generator for DSP. http://www.xilinx.com/tools/sysgen.htm
207. Altera (2011) DSP builder/simulink. http://www.altera.com/technology/dsp/dsp-builder/dsp-simulink.html
208. Mathworks (2011) Simulink HDL coder. http://www.mathworks.com/products/slhdlcoder/
209. Synopsys (2011) High-level synthesis with synphony model compiler. http://www.synopsys.com/Systems/BlockDesign/HLS/Pages/Synphony-Model-Compiler.aspx
210. Zoss R, Habegger A, Bandi V, Goette J, Jacomet M (2011) Comparing signal processing hardware-synthesis methods based on the matlab tool-chain. pp 281–286
211. Levine DL, Flores-Gaitan S, Gill CD, Schmidt DC (1999) Measuring OS support for real-time CORBA ORBs. In: Object-oriented real-time dependable systems. Proceedings, 4th international workshop on 1999, pp 9–17
212. Wind River (2011) Workbench. http://www.windriver.com/products/workbench/
213. TenAsys Corporation (2011) INtime for windows. http://www.tenasys.com/products/intime.php
214. Lynuxworks (2011) LynxOS RTOS. http://www.lynuxworks.com/rtos/rtos.php
215. ATLAS (2011) Automatically tuned linear algebra software. http://math-atlas.sourceforge.net/
216. Yang X (2004) A feasibility study of UML in the software defined radio. In: Electronic design, test and applications. DELTA 2004, 2nd IEEE international workshop on 2004, pp 157–162
217. Dohler M, Heath RW, Lozano A, Papadias CB, Valenzuela RA (2011) Is the PHY layer dead? Commun Mag, IEEE 49:159–165
218. Rappaport TS (2002) Wireless communications: principles and practice. Prentice Hall PTR, Englewood Cliffs
219. Haykin M (2009) Communication systems. Wiley, New York
220. Reed JH (2002) Software radio: a modern approach to radio engineering. Prentice Hall, Englewood Cliffs
221. Johnson WA (2004) Telecommunication breakdown: concepts of communication transmitted via software-defined radio. Pearson Education Inc, New York
222. Kenington PB (2005) RF and baseband techniques for software defined radio. Artech House, London
223. Bhatnagar MR, Hjorungnes A (2007) SER expressions for double differential modulation. In: Information theory for wireless networks, IEEE information theory workshop on 2007, pp 1–5
224. Shiu D-S, Kahn JM (1999) Differential pulse-position modulation for power-efficient optical. Commun, IEEE Trans 47:1201–1210
225. Park H, Barry JR (2004) Trellis-coded multiple-pulse-position modulation for wireless infrared communications. Commun, IEEE Trans 52:643–651
226. Aiello GR, Rogerson GD (2003) Ultra-wideband wireless systems. Microwave Mag, IEEE 4:36–47
227. Lin Y-P, Phoong S-M, Vaidyanathan PP (2010) Filter bank transceivers for Ofdm and Dmt systems. Cambridge University Press, Cambridge
228. Waldhauser DS, Nossek JA (2006) Multicarrier systems and filter banks. Advances in radio science. http://www.adv-radio-sci.net/4/165/2006/ars-4-165-2006.pdf

229. Goldsmith Andrea (2005) Wireless communications. Cambridge University Press, Cambridge
230. Welch L (1974) Lower bounds on the maximum cross correlation of signals. Inf Theory, IEEE Trans 20:397–399
231. Massey JL, Mittelholzer T (1991) Welchs bound and sequence sets for code-division multiple-access systems, pp 63–78
232. Jayasimha S, Paladugula J (2008) Interference cancellation in satcom and ranging. In: National conference on communications
233. Comtech EF Data, DoubleTalk carrier-in-carrier performance characterization. http://www. appsig.com/documentation/dt_cnc_performance.pdf
234. Foschini GJ (1996) Layered space–time architecture for wireless communication in a fading environment when using multielement antennas. Bell Labs Tech J 41–59
235. Telatar E (1995) Capacity of multiantenna Gaussian channels. AT&T, Tech Memo
236. Arapoglou P-D, Burzigotti P, Bertinelli M, Bolea-Alamanac A, De Gaudenzi R (2011) To MIMO or not to MIMO in mobile satellite broadcasting systems. Wirel Commun, IEEE Trans 10:2807–2811
237. Huang KD, Malone D, Duffy KR (2011) The 802.11 g 11 Mb/s rate is more robust than 6 Mb/s. Wirel Commun, IEEE Trans 10:1015–1020
238. Parsons J (2000) The mobile radio propagation channel. Wiley, New York
239. Sklar B (1997) Rayleigh fading channels in mobile digital communication systems. Commun Mag, IEEE 35:90–100
240. Crane RK (1977) Ionospheric scintillation. Proc IEEE 65:180–199
241. Moulsley T, Vilar E (1982) Experimental and theoretical statistics of microwave amplitude scintillations on satellite down-links. Antennas Propag, IEEE Trans 30:1099–1106
242. Filip M, Vilar E (1990) Adaptive modulation as a fade countermeasure. An olympus experiment. Int J Satell Commun 8:31–41
243. Olsen R, Rogers D, Hodge D (1978) The aR^b relation in the calculation of rain attenuation. Antennas Propag, IEEE Trans 26:318–329
244. Lin D-P, Chen H-Y (2002) An empirical formula for the prediction of rain attenuation. Antennas Propag, IEEE Trans 50:545–551
245. Willis MJ Rain effects. http://www.mike-willis.com/Tutorial/PF10.htm
246. Willis MJ (1991) Fade countermeasures applied to transmissions at 20/30 GHz. Electron Commun Eng J 3:88–96
247. Amin MG, Wang C, Lindsey AR (1999) Optimum interference excision in spread spectrum communications. Sig Process, IEEE Trans 47:1966–1976
248. Ouyang X, Amin MG (2001) Short-time fourier transform receiver for nonstationary. Sig Process, IEEE Trans 49:851–863
249. Tarasov KN, McDonald EJ, Grayver E (2008) Power amplifier digital predistortion—fixed or adaptive? In: Military communications conference, MILCOM 2008, IEEE, pp 1–7
250. Nezami MK (2002) Performance assessment of baseband algorithms for direct conversion tactical software defined receivers: I/Q imbalance correction, image rejection, DC removal, and channelization. In: MILCOM 2002, Proceedings, vol 1. pp 369–376
251. Mengali U (1997) Synchronization techniques for digital receivers. Plenum Press, New York
252. Rutman J, Walls FL (1991) Characterization of frequency stability in precision frequency. Proc IEEE 79:952–960
253. Ferre-Pikal ES et al (1997) Draft revision of IEEE STD 1139–1988 standard definitions of. In: Frequency control symposium. Proceedings of the 1997 IEEE international, pp 338–357
254. Hogg J, Bordeleau F (2010) SCA deployment management: bridging the gap in SCA development. http://www.zeligsoft.com/files/whitepapers/SCA-Deployment-Management-WP.pdf
255. Fast Infoset http://en.wikipedia.org/wiki/Fast_Infoset
256. Bayer M (2005) Analysis of binary XML suitability for NATO tactical messaging. Naval Postgraduate School, Masters thesis, ADA439540

257. XML Query http://www.w3.org/XML/Query/
258. Bard J (2007) Software defined radio. Wiley, New York
259. Microsoft (2011) Microsoft research software radio (Sora). http://research.microsoft.com/en-us/projects/sora/
260. Texas Instruments (2011) ADS1675: 4MSPS, 24-Bit analog-to-digital converter. http://focus.ti.com/docs/prod/folders/print/ads1675.html
261. Xilinx (2009) Xilinx Virtex-4 configuration guide. www.xilinx.com/support/documentation/user_guides/ug071.pdf
262. Clauss RC (2008) Cryogenic refrigeration systems. In: Low-noise systems in the deep space network. Jet Propulsion Laboratory, Pasadena, Chap. 4, p 161
263. Xilinx (2010) Radiation-hardened, space-grade Virtex-5QV device overview. http://www.xilinx.com/support/documentation/data_sheets/ds192_V5QV_Device_Overview.pdf
264. iBiquity, HD Radio. http://www.hdradio.com/
265. World DAB, Digital multimedia broadcasting. http://www.worlddab.org/
266. Neo WC et al (2006) Adaptive multi-band multi-mode power amplifier using integrated varactor-based tunable matching networks. Solid-State Circuits, IEEE J 41:2166–2176
267. Yumin Lu, Peroulis D, Mohammadi S, Katehi LP (2003) A MEMS reconfigurable matching network for a class AB amplifier. Microwave Wirel Compon Lett, IEEE 13:437–439
268. Hadzic I, Udani S, Smith J (1999) FPGA viruses. In: 9th international workshop on field-programmable logic and applications, Glasgow, UK
269. Chen SY, Kim N-S, Rabaey JM (2008) Multi-mode sub-Nyquist rate digital-to-analog conversion for direct waveform synthesis. In: Signal processing systems, 2008. SiPS 2008, IEEE workshop on 2008. pp 112–117
270. Murmann B (2008) A/D converter trends: power dissipation, scaling and digitally assisted architectures. In: Custom integrated circuits conference, CICC 2008. IEEE, pp 105–112
271. Courreges S et al (2010) Back-to-back tunable ferroelectric resonator filters on flexible organic substrates. Ultrason, Ferroelectr Freq Control, IEEE Trans 57:1267–1275
272. Alexeff I et al (2005) Advances in plasma antenna design. In: Plasma science, ICOPS '05. IEEE conference record—abstracts. IEEE international conference on 2005. pp 350–350
273. Dominic Tavassoli (2011) Ten steps to better requirements management. http://public.dhe.ibm.com/common/ssi/ecm/en/raw14059usen/RAW14059USEN.PDF
274. Gaudet VC, Gulak PG (2003) A 13.3-Mb/s 0.35 µm CMOS analog turbo decoder IC with a configurable interleaver. Solid-State Circuits, IEEE J 38:2010–2015
275. Huang J, Meyn S, Medard M (2006) Error exponents for channel coding with application to signal constellation design. Sel Areas Commun, IEEE J 24:1647–1661
276. Sayegh S (1985) A condition for optimality of two-dimensional signal constellations. Commun, IEEE Trans 33:1220–1222
277. Foschini GJ, Golden GD, Valenzuela RA, Wolniansky PW (1999) Simplified processing for high spectral efficiency wireless. Sel Areas Commun, IEEE J 17:1841–1852
278. Mitola J (2000) Cognitive radio: an integrated agent architecture for software defined radio. Royal Institute of Technology (KTH), Kista, Sweden, PhD thesis, ISSN 1403-5286
279. Jensen K, Weldon J, Garcia H, Zettl A (2007) Nanotube radio. Nano Lett 7:3508–3511
280. Bonello N, Yang Y (2011) Myths and realities of rateless coding. IEEE Commun Mag 9(8):143–151
281. Cavallaro JR, Vaya M (2003) Viturbo: a reconfigurable architecture for Viterbi and turbo decoding. In: Acoustics, speech, and signal processing, Proceedings, (ICASSP '03). IEEE international conference on 2003, vol 2. pp II–497–500
282. Xilinx (2011) Xilinx partial reconfiguration. http://www.xilinx.com/tools/partial-reconfiguration.htm
283. Altera (2011) Stratix V FPGAs: ultimate flexibility through partial and dynamic reconfiguration. http://www.altera.com/products/devices/stratix-fpgas/stratix-v/overview/partial-reconfiguration/stxv-part-reconfig.html

284. Reed JH (2005) An introduction to ultra wideband communication systems. Prentice Hall PTR, Englewood Cliffs

285. Wang H, Guo J, Wang Z (2007) Feasibility assessment of repeater jamming technique for DSSS. In: Wireless communications and networking conference, WCNC 2007. IEEE, pp 2322–2327

286. Tabula (2011) Spacetime 3D architecture. http://www.tabula.com/technology/technology.php

287. Rondeau T (2011) VOLK and GNU radio. In: NEWSDR, Boston, http://people.bu.edu/mrahaim/NEWSDR/Presentations/NEWSDR_Rondeau2.pdf

288. Techma (2011) NeXtMidas. http://nextmidas.techma.com/

289. Kostina V, Loyka S (2011) Optimum power and rate allocation for coded V-BLAST: instantaneous optimization. Commun, IEEE Trans 59:2841–2850

290. Knuth D (1974) Structured programming with go to statements. ACM J Comput Surv 6(4):268

291. Fog A (2011) Software optimization resources. http://www.agner.org/optimize/

292. Shrestha R, Mensink E (2006) A wideband flexible power upconverter for software defined radios. pp 85–89

293. Simon A (2002) The COT planning guide: tips, tactics, and strategies for successful IC outsourcing. Simon Publications, Bethesda

Index

A
A/D converters, 7
Adaptive coding and
 modulation, 10, 30
Adaptive equalization, 232
Additive white
 Gaussian noise, 229
Adjacent channel interference, 203
Analog radio, 1
Antenna, 1, 92, 128, 157
 adaptive 180
 diversity, 235
ARQ, 30, 32
ASIC, 37, 51, 84, 188

B
Bandwidth efficiency, 10, 12, 14, 206,
 224, 228
Beam pattern, 178
Beamforming, 178, 221
Blanking, 238

C
Carrier frequency, 7, 21, 22, 94, 109, 152, 153,
 160, 179, 203
Channel, 71
Channel emulator, 31, 70, 236, 243
Channel selection, 7
Chip, 214
Coding gain, 228
Cognitive radio, 9, 21, 25, 65
 cognition engine, 23
Coherence time, 236

Connectorized, 166
Constellation, 209
Cost, 2, 20, 27, 37, 47, 54, 95, 109, 123, 136,
 137, 140, 148, 159, 166, 190, 194, 245
Cyclic prefix, 213, 233

D
Datapath, 44, 47, 75, 190
DBRA, 17
Desirements, 160
Direct rf synthesis, 152
Direct-if, 155, 165, 239
Diversity, 221
Doppler, 28, 240
DSP, 2, 43
DSSS, 53, 57, 64, 213, 214, 235, 238
 asynchronous, 214
DVB-S2, 14, 118

E
Eb/N0, 224
Effective number of bits, 167
Error correcting code, 226
Error floor, 232
Error-vector-magnitude, 238
Event driven, 136
Excision, 237

F
Fast Fading, 13, 233
FCC, 10, 21, 22, 204

E. Grayver, *Implementing Software Defined Radio*,
DOI: 10.1007/978-1-4419-9332-8, © Springer Science+Business Media New York 2013

F (*cont.*)
Feedback loop, 240
Feedforward estimator, 240
FHSS, 238
Firmware, 61
FPGA, 16, 38, 44, 54, 60, 73, 84, 140, 183,
    188, 190
Frequency shift keying, 208

G
GNU Radio, 69, 131
    GRC, 188
GPP, 55, 73
GPU, 47
Green radio, 25
Guard time, 216

H
Handover, 20
Hardware acceleration, 53, 65, 84
    FPGA, 91
Hidden node problem, 23

I
Incumbent user, 22
Intel IPP, 138
Interference, 29
Intermodulation, 8, 238
IQ mixer, 154
Iterative decoding, 227

J
Jammer, 29, 238

M
Maximal ratio combining, 235
Modem hardware abstraction layer, 101, 107
Modulation, 207
Modulation index, 208
Multi user detection, 238
Multipath, 230
Multiuser detection, 24

N
Nonlinear distortion, 238

O
Object Request Broker, 100
OFDM, 17, 233, 240
On-off-keying, 207
Opportunistic beamforming, 27
OSSIE, 100, 117, 131, 140, 194, 245

P
Packet, 16, 73
Parity, 3, 126, 226
Partial reconfiguration, 17, 60, 62, 68
Performance loss, 225
Phase noise, 240
Physical layer, 5
Power consumption, 5, 10, 24, 27, 38, 48, 52,
    55, 81, 154, 158, 165, 172
Primary user, 21
Profiling, 85, 195

Q
Q factor, 174
Q-factor, 157
Quadrature IF mixer, 154

R
Rateless codes, 226
Rayleigh model, 236
RF front end, 2, 38, 145
RF front ends, 118

S
Shannon limit, 10, 26, 199, 205
Signal acquisition, 3, 186, 57, 61, 138, 65
Signal processing workstation (SPW), 187
Simulink, 187
Skeleton, 100
Software controlled, 6
Spatial multiplexing, 222
Spectrum sensing, 22
SPU, 47
STRS, 97, 108
Stub, 100
Symbol, 202
Symbol rate, 208
Symbols, 207
Synchronization, 240
SystemVue, 187, 192

T
Testing, 39
Tilera, 49
Time slot, 216
Tradeoffs, 8

V
VITA-49, 93, 94, 118, 122, 123, 145, 146, 149
VRT. See VITA-49

W
Waterfall curves, 224
Waterfilling, 223
Waveform, 5
Wireless innovation forum, 5

Z
Zero-if, 239

CPSIA information can be obtained
at www.ICGtesting.com
Printed in the USA
LVOW13*0925090717

540733LV00005B/283/P